THE CLOCKWORK FACTORY

Women and Work in Fascist Italy

PERRY R. WILLSON

CLARENDON PRESS · OXFORD
1993

Oxford University Press, Walton Street, Oxford OX2 6DP
Oxford New York Toronto
Delhi Bombay Calcutta Madras Karachi
Kuala Lumpur Singapore Hong Kong Tokyo
Nairobi Dar es Salaam Cape Town
Melbourne Auckland Madrid
and associated companies in
Berlin Ibadan

Oxford is a trade mark of Oxford University Press

Published in the United States
by Oxford University Press Inc., New York

British Library Cataloguing in Publication Data
Data available

Library of Congress Cataloging in Publication Data
Willson, Perry R.
The clockwork factory: women and work in Fascist Italy/Perry R. Willson.
p. cm
Includes bibliographical references and index.
1. Women electric industry workers—Italy—Milan—History—20th century. 2. Magneti Marelli (Firm)—History. I. Title.
HD6073.E321818 1993 93–14160
331.4'8213'094521—dc20
ISBN 0–19–822732–9

1 3 5 7 9 10 8 6 4 2

Typeset by Cambrian Typesetters
Frimley, Surrey
Printed in Great Britain
on acid-free paper by
Biddles Ltd., Guildford, Surrey

THE CLOCKWORK FACTORY

ACKNOWLEDGEMENTS

MANY people have helped me during the preparation of this book. In particular I should like to thank, for their various help and support in Italy, the staff of the Magneti Marelli Personnel Department (and in particular Dottoressa Bossi) for granting me access to the firm's archives; Duccio Bigazzi for a great deal of encouragement and advice; Luigi Ganapini for his assistance with getting some of my interviews transcribed; Egeo Mantovani for his energetic efforts to track down ex-Magneti workers for me to interview; Rosaria Moccia for her help with the initial interviews; Luisa Passerini and Chiara Saraceno, who advised me on drawing up my questionnaire; Paolo Natale for computing assistance; and Gianfranco Petrillo and Giuseppe Vignati, both of ISRMO in Sesto San Giovanni, for their generous help and advice on numerous occasions. I am also grateful to the Fondazione Einaudi (Turin) for their financial support and to the Fondazione Feltrinelli (Milan) for the use of their excellent library facilities. In Britain, I should like to thank Stuart Woolf for his guidance, Diane Benn for her help, advice, and tolerance in putting up with a writer's crises, and Ludmilla Jordanova, Steve Smith, and Jonathan Steinberg for their constructive comments on an earlier draft.

Last but not least, a special word of thanks is due to all the former Magneti Marelli workers and clerks who gave up their time to share some of their past with me.

P. R. W.

Edinburgh
1991

CONTENTS

LIST OF TABLES

Introduction

INTER-WAR Italy, like any other modern society, was shaped by the relations of power between the sexes. This fundamental aspect of the Fascist experience has been sorely neglected by many historians. Despite the fact, only recently beginning to be recognized, that gender is one of the prime forces in historical change, much of the enormous volume of research which has been published on Italian Fascism is 'gender blind'. Most histories of the Fascist period have been impoverished by the fact that they have given no more than a cursory reference to the role of the female half of the inter-war Italian population.

Since the upsurge in interest in women's history in the late 1970s following the re-emergence of the feminist movement, this situation has improved to some extent. A certain amount of relevant research has been done, including studies of Fascist ideology, changes in the family, women's contribution to the Resistance, and the Fascist women's organizations. Some good work has been published, but there is still a need for carefully theorized work on many aspects of women's history in the Fascist period. Furthermore, most of the historical writing not explicitly directed towards a study of women continues simply to ignore them. Women's history remains marginalized and much of the task of 'gendering' the writing of Italy's modern history remains to be done.

Historians' neglect of this aspect of the Fascist experience in Italy is perhaps particularly striking given that one of the key elements which lay at the heart of Fascist ideology was a belief in starkly contrasting gender roles. A great deal of attention was devoted to the 'woman question' by Mussolini and his followers, who evolved a definite, albeit frequently contradictory, ideology about women's place in society and directed specific policies towards this half of the population.[1]

[1] On Fascist ideology towards women see e.g. P. Meldini, *Sposa e madre esemplare: ideologia e politica della donna e della famiglia durante il fascismo* (Rimini–Florence, 1975); E. Santarelli, 'Il fascismo e le ideologie antifemministe', *Problemi*

WOMEN IN FASCIST IDEOLOGY. 'ANGELS OF THE HEARTH'

Early Fascism did not seem explicitly antifeminist and even called for female suffrage, but by the mid-1920s official ideology became increasingly misogynous. Gender roles were defined in terms of a crude biological determinism, any form of emancipation for women was opposed,[2] and motherhood was extolled as the only real contribution women could make to society. This Fascist ideal of woman-as-mother was expounded in a series of demagogic speeches, radio broadcasts, newspaper articles, and stage-managed pageantry, and a deluge of prescriptive writings appeared detailing ideal female behaviour.[3] Women were exhorted to stay at home (where they were referred to as 'angels of the hearth'), and, above all, make sacrifices in the cause of their country by producing plenty of babies essentially as cannon fodder for the imperialist drive. Women's waged employment outside the home was denounced from the 1920s, and with the economic crisis and growing unemployment this campaign became more vehement. In 1934, for example, writing in *Il Popolo d'Italia*, Mussolini declared that mechanization and women's work were the main causes of the scarcity of jobs for men. For women, he asserted, work was potentially dangerous and might make them sterile, whereas for men it was a source of 'great physical and moral virility'.[4]

It seems reasonable to argue that much of such rhetoric was in fact partly aimed at the Church or at male Italians, to reassure them that their power would not be undermined. None the less it was also directed at women themselves in an attempt to make

del socialismo (1976); E. P. Noether, 'Italian Women and Fascism: A Re-evaluation', *Italian Quarterly*, 90 (1982); M. A. Macciocchi, *La donna nera: consenso femminile e fascismo* (Milan, 1976); Macciocchi (ed.), *Les Femmes et leurs maîtres* (Paris, 1978); Macciocchi, 'Female Sexuality in Fascist Ideology', *Feminist Review*, 1 (1979). See also Victoria De Grazia's important new study *How Fascism Ruled Women* which is due to be published in 1992 by University of California Press. Unfortunately I have been unable to take account of this work as it has not yet appeared at the time of writing.

[2] In May 1925 some women were granted the right to vote in local elections, but this was purely an empty gesture given that local elections were abolished the following year.

[3] For extracts from and an analysis of some of these writings the best source is Meldini, *Sposa e madre esemplare*.

[4] B. Mussolini, 'Macchina e donna', *Il Popolo d'Italia*, 206 (31 Aug. 1934).

them feel rewarded for accepting their allotted role in the Fascist state. Thus the sacrifices that women were supposed to make were romanticized, and the reduction of women to mere breeding machines without intellect was raised up into a great mission to provide colonists for the empire. Widows of war heroes and 'prolific mothers' were given heroic status. These rather crude attempts to mystify women's role were accompanied by grandiose ceremonies where, for example, women were given medals for producing large families or were asked to donate their wedding rings for the Ethiopian war effort.

Pro-natalist legislation was also introduced to encourage women to dedicate their lives to motherhood. This was done mainly through the promotion of early marriage and the outlawing of abortion, contraception, and even contraceptive information. Early marriage was fostered by a tax on bachelors[5] and the introduction of some other (fairly limited) incentives such as employment preference systems and cheap housing for married men. Other provisions gradually introduced included special benefits for large families, marriage loans, and family allowances.[6]

With such a relentless emphasis on women's reproductive role to the detriment of any other, Fascist ideology about gender roles was undeniably particularly misogynous. However, it is also important to recognize, as Emiliana Noether has argued, that not all these ideas were new:

> If the Fascist years are considered to be part of a historical continuum rather than an isolated episode, the perspective shifts considerably. Many of the repressive aspects of the late 1920s and '30s become overt reflections of traditional Italian views on women's place in the political, economic, and social structure of society.[7]

[5] This tax (introduced in 1926) was levied on all unmarried men between the ages of 25 and 65. It was in two parts: a flat rate to be paid annually and also a 25% tax on total income.

[6] Most texts on women and fascism give some coverage of the demographic campaign. Perhaps the most useful in English is the excellent article by L. Caldwell, 'Reproducers of the Nation: Women and the Family in Fascist Policy', in D. Forgacs (ed.), *Rethinking Italian Fascism* (London, 1986). Caldwell makes the important point that many modern European countries introduced these sorts of policies in the 20th cent. It was, therefore, less the initiatives themselves which were specifically 'Fascist' in nature than the fanfare of ideology preaching women's subordination and the primacy of motherhood.

[7] Noether, 'Italian Women and Fascism', 69.

One way in which this 'continuum' is evident is in the attitude of the Catholic Church. Reconciled to the state with the Lateran Accords of 1929, the Church continued to be a highly important social and political force in this period.[8] In many respects Catholic and Fascist ideas about male and female roles in society were very similar and the Vatican had little difficulty in accepting Fascist ideology in this respect. Women were faced with broadly the same arguments from pulpit and politicians alike.[9] According to the Church, women were biologically inferior to men and the two sexes should remain in 'separate spheres' of activity. Good Catholic women were to stay at home devoting themselves to motherhood, and work outside the home was frowned upon.

Mussolini's ideas on the role of women were, therefore, not totally original. Although some advances had been made for Italian women during the Liberal period,[10] pre-Fascist Italy was far from an emancipated paradise and most Italian women had been hearing similar speeches from their local priests before the arrival of the Fascist Blackshirts. Where Church and State differed was in the stress placed by Mussolini on the imperialist role of the 'prolific mothers', whereas the Church's primary concern was with preserving the 'virtue' and 'respectability' of women who might be tainted by contact with the outside world.

Fascist ideology on the role of women is fairly easy to document. Harder to evaluate is the impact of these ideas. It is well known, for example, that the pro-natalist initiatives had virtually no discernible impact on the demographic statistics of inter-war Italy, but the extent to which they shaped women's

[8] For a history of the relations between Church and State in this period see e.g. P. Scoppola, *La chiesa e il fascismo: documenti e interpretazioni* (Bari, 1971); D. A. Binchy, *Church and State in Fascist Italy* (Oxford, 1941); C. A. Jemolo, *Church and State in Italy, 1850–1950* (Oxford, 1960); J. Pollard, *The Vatican and Italian Fascism 1929–1932: A Study in Conflict* (Cambridge, 1985).

[9] On the similarities and differences between Catholic and Fascist ideas about women see Macciocchi, *La donna nera*; Meldini, *Sposa e madre esemplare*; and Caldwell, 'Reproducers of the Nation'.

[10] e.g. there had been some very limited improvements in the legal position of women in the immediate post-war period, such as the opening up of some new areas of the civil service to women and granting them the right to practise law. But even these small advances were hedged around with restrictions and they were still excluded from certain civil service jobs, particularly higher-ranking ones.

self-image is not so easy to assess. Furthermore, different groups of women were affected in different ways. Fascism may have expounded one universal ideal of true womanhood, but the female experience in this period was, of course, varied and complex. There was less a single identifiable 'women's experience of Fascism' than a plurality of experiences dependent upon a number of factors. Women were far from a homogenous group, and as Chiara Saraceno has quite rightly noted,

> it is too simplistic to refer to 'women' and even 'working class women' is too broad and abstract a category, since there are great differences—and these were even greater in Fascist Italy—between the social classes and even within the working class itself, according to geographical location, economic sector, cultural factors, and generation.[11]

WOMEN AND WORK

In practice, as employment policy shows, the Fascists themselves were fully aware of such distinctions and viewed women as a single uniform category only in rhetorical speeches. Legislation tended to treat separate groups of women in different ways and policy was mostly class specific.

Despite their endless rhetoric, the Fascist regime never attempted to compel working-class women to return to home and hearth by banishing them from the labour force. At least some Fascists recognized that even where households contained a male breadwinner, in the absence of the 'family wage', the aim of the demographic campaign to make women full-time mothers was unrealistic for most working-class and peasant families. Furthermore, female low wages were simply too profitable to Fascism's capitalist backers for the regime to contemplate really excluding women from the industrial and agricultural workforce. None the less the state did intervene 'protectively' in their world of work, with the declared aim of offsetting the alleged negative impact that work could have on women's reproductive capacities.

Ironically some of this meant changes which could actually

[11] C. Saraceno, 'Percorsi di vita femminile nella classe operaia. Tra famiglia e lavoro durante il fascismo', *Memoria*, 2 (1981).

make it easier for women to work outside the home, such as improvements in maternity leave and the provision of nurseries and breast-feeding rooms. Other legislation was more restrictive, such as the law of April 1934 which regulated female and child labour.[12] Under this law women were forbidden to do night and underground mining work, limitations were placed on the number of hours they could work,[13] and differential scales were established of the maximum weights that male and female workers were allowed to carry. This law also created a type of extended childhood for women by establishing a category of 'female minors' (those over 15 but under 21 years of age), who, together with children of either sex under 15,[14] were banned from certain types of jobs which were said to be heavy, dangerous, or dirty.[15]

The impact of such measures is hard to judge. They were probably widely ignored. Moreover, vast numbers of working women were excluded. The legislation did not extend to, amongst others, domestic servants, members of the employer's family, persons working at home, and most agricultural workers.[16] For the Fascist legislators, work that was dangerous to a woman's health in a factory was apparently not so if she was working at home or for a member of her own family. Such exclusions suggest that the legislators were at least as worried by female extradomestic work in factories as by the unhealthiness of particular jobs. Significantly, 'female minors' were in precisely the age group which was most likely to be employed outside the home, in factories for example.

Furthermore, in the light of Fascist ideology which emphasized female inferiority and the primacy of motherhood, it is impossible to view protective legislation as purely and simply in

[12] Law no. 653, 26 Apr. 1934.

[13] Women were not allowed to work more than 11 hours in one shift and not more than 6 hours without a break.

[14] For most jobs the minimum age at which a child of either sex could be employed was 14, although sometimes this could be reduced to 13 or even 12.

[15] The regulations accompanying this law, published two years later, gave a long list of jobs. The list was divided into two sections. One was of jobs which 'female minors' or boys under 15 were banned from doing, and the other of jobs which they could do only if special health and safety precautions were taken. See Royal Decree (R.d.) no. 1720, 7 Aug. 1936.

[16] These categories were also excluded from most of the improvements in maternity protection. See R.d. no. 654, 22 March 1934.

women's best interests. Even ostensibly 'protective' laws may have been more concerned with protecting male status and privilege than protecting the women themselves. Such legislation reinforced the idea that, even though women's work was tolerated in practice, it was for their reproductive potential that they were really valued. It also made women's labour slightly more expensive, gave men a monopoly in certain (if mainly fairly undesirable) types of work, and helped reinforce the idea that certain jobs were inherently 'men's work' and others were 'women's'. All this may have limited levels of female employment in some fields.

Fascism did, therefore, create some obstacles to working-class women's access to the waged labour market.[17] A much more sustained attack, however, was directed at middle-class women. Here too policies were legitimated by invoking the need to raise the birth rate. But this was far from the sole motivation. Legislation for middle-class women had more teeth, because the desire to safeguard the power and privilege of male white-collar workers and professionals was perceived as particularly important. By such laws the Fascists aimed to shore up the position of the lower-middle-class male (a key component of the mass base of the regime) whose position in the family and workplace appeared threatened by the rise of 'white blouse' employment and the (admittedly somewhat weak) feminist movement of the turn of the century.[18] Restrictions on female employment appealed to such men because of their own insecurity in the labour market, an insecurity which stemmed from Italy's endemic 'intellectual unemployment'. Even without female competition the numbers of applicants persistently far out-

[17] This somewhat undermines De Grand's rather dismissive claim that gender-specific Fascist employment legislation targeted only the middle class, arguing that the 'suppression of peasant and proletarian women formed a particular aspect of the general repression of the entire working class': A. De Grand, 'Women under Italian Fascism', *Historical Journal*, 4 (1976), 947.

[18] There have been quite a number of studies of the turn of the century feminist movement. See, for example: M. Casalini, 'Femminismo e socialismo in Anna Kuliscioff 1890–1907', *Italia contemporanea*, 143 (1981); F. Pieroni Bortolotti, *Socialismo e questione femminile in Italia, 1892–1922* (Milan, 1974); *Anna Kuliscioff e l'età del riformismo* (Milan, 1977); S. Puccini, 'Condizione della donna e questione femminile (1892–1922)' in G. Ascoli (ed.), *La questione femminile in Italia dal '900 ad oggi* (Milan, 1979); A. R. Buttafuoco, *Le Mariuccine* (Milan, 1985).

stripped the supply of posts in state employment and the professions.[19]

It was not long after the Fascists' rise to power that laws restricting middle-class women's employment opportunities began to be passed. The first main target sector was education. In 1923 women were banned from headships of middle schools.[20] Three years later[21] the recruitment of female secondary school teachers in specific subjects such as literature, philosophy, Latin, Greek, and history was blocked. In the 1930s the scope of the legislation was widened and from 1933 all branches of the State administration were authorized to exclude women from competitive examinations for new civil service posts or place a ceiling on their numbers.[22]

Particularly notorious is the decree law of 1938 which limited the recruitment of women in both private and public sector employment to 10 per cent. In the case of small businesses no female clerks at all were to be recruited.[23] This law has been seen as an extremely vehement attack on women's right to work. Noether, for example, asserts[24] that this was the first time Fascist legislation extended its attack on female employment from the middle classes to employed women in general. However, discriminatory and offensive as it undoubtedly was to women, this law was far less drastic than some historians have suggested.[25] The total ban on female clerks in small businesses

[19] This was created largely by the peculiarities of Italy's education system combined with the fact that the slow pace of industrialization hindered the expansion of private sector employment. On this problem see M. Barbagli, *Disoccupazione intellettuale e sistema scolastico in Italia, 1859–1973* (Bologna, 1974).

[20] R.d. no. 1054, 6 May 1923, art. 12. This was extended to private middle schools in 1925 (R.d. no. 1084, 6 June 1925).

[21] R.d. no. 2480, 9 Dec. 1926.

[22] R.d. no. 1554, 28 Nov. 1933.

[23] R.d. no. 1514, 5 Sept. 1938. This law, however, came to nothing, as it was to be implemented over a period of three years, and a mere nine days after Italy's entry into the Second World War the part of the legislation dealing with public sector employment was suspended indefinitely (Law no. 739, 29 June 1940).

[24] 'Until the 1938 decree, efforts to limit the number of working women had been aimed mainly at the middle class' (Noether, 'Italian Women and Fascism', 73).

[25] De Grand e.g. describes it as a 'draconian measure' ('Women under Italian Fascism', 965). Noether calls it 'the notorious decree law . . . which drastically curtailed the employment of women' ('Italian Women and Fascism', 73).

was considerably softened by the fact that it made an exception for family members. Perhaps more importantly, despite the somewhat ambiguous wording, it is fairly clear that the quotas were to be implemented only in 'middle-class jobs'. One especially revealing clause demonstrates that the 1938 law was never intended to expel women from the waged labour force. This clause established that the quotas did not apply to jobs seen as 'particularly suitable for women'. The list of non-quota jobs, issued the following year,[26] included many of the main jobs which female clerical workers actually did, generally the more menial and least prestigious ones. No quotas were applied, for example, on the numbers of shorthand typists and telephonists. The 10 per cent ceiling, therefore, was concerned primarily with preserving men's ability to monopolize the higher-status jobs by confining women to certain work roles and checking the expansion of female employment in specific types of white-collar and professional work.

In view of the nature of this and the other restrictive laws which preceded it, it is perhaps unsurprising that they had little or no discernible effect on the overall numbers of women in waged employment. According to census figures large numbers of women continued to participate in the labour force throughout the inter-war period, engaged in a variety of occupations in many different contexts.[27] Admittedly official statistics do show a slight decline in the numbers of women working between 1921 and 1936,[28] but it would be wrong to assume that this was caused primarily, if at all, by Fascist campaigns against female employment. This decrease was not specific to the Fascist period but part of a longer-term trend which had begun long before the rise of the Blackshirts and continued after their fall. Fascist policy may have had some part to play in encouraging the

[26] R.d. no. 898, 29 June 1939.

[27] e.g. the 1936 census recorded that about 25% of the industrial labour force was female (1,377,373 women workers). The main sectors in which they were to be found were textiles (75% of the labour force), paper (50.7%), leather (40.8%), chemicals, food processing, mechanical industries, and printing.

[28] Within this period there were variations. Between 1921 and 1931 male employment through Italy rose whilst female employment fell. Conversely in the period 1931–6 women's employment rose while men's fell, probably due to women taking over some men's jobs in the Ethiopian War. See P. Sabbatucci Severini and A. Trento, 'Alcuni cenni sul mercato di lavoro durante il fascismo', *Quaderni storici*, 29–30 (1975).

continuation of this trend but other factors, such as the decline in the textile industry, were probably more important.

Even in the case of middle-class women's employment, more explicitly targeted by Fascist legislation, there is no real evidence that numbers were significantly affected. In the tertiary sector, apart from the areas of public administration directly affected by the legislation, female employment actually rose. Even in education the impact appears small. Many women continued to teach, and the numbers of women attending university, gateway to a teaching career or the professions, steadily increased.[29]

Census figures, of course, present many problems to the historian and should be treated as no more than a rough guide. An alternative explanation for this apparent fall in female employment may quite simply be that these statistics are unreliable.[30] Given that during her lifetime one woman might do a succession of different full- or part-time jobs according to age and marital or parental status, the way in which she described herself on a census form might be of limited significance. Many of those counted as housewives may have actually been doing a certain amount of paid employment which had escaped unrecorded due to its intermittent, part-time, or domestic nature. As such the decline in the official figures may in fact simply indicate a rise in the numbers of women in marginal, temporary, or seasonal work.

The question of work, central to an examination of the role of women in the Fascist period, is far from exhausted by a look at

[29] The numbers of both male and female students rose, but the greatest rise was amongst women. In 1911–12 only 3% of university students were female. By 1935–6 this had risen to 17.4% and by 1942–3 women made up almost a third (29.9%) of the university population (De Grand, 'Women under Italian Fascism', 960). Female students were to be found in most faculties but the largest numbers opted for philosophy and humanities. In teacher-training institutes there were also rising numbers of women.

[30] On this problem in the context of 19th-century England and Wales see E. Higgs, 'Women, Occupations and Work in the 19th Century Censuses', *History Workshop*, 23 (1987). For a similar study of Rome in the 1860s, see the excellent article by M. Pelaia, 'Mestieri femminili e luoghi comuni. Le domestiche a Roma a metà Ottocento', *Quaderni storici*, 68 (1988). In inter-war Italy even the census compilers themselves admitted that their figures were far from accurate when it came to women. A note in the report on the 1931 census warned that the compilers felt that it was impossible to claim great accuracy with regard to female employment in agriculture, commerce, and crafts. (See Noether, 'Italian Women and Fascism', 72.)

census figures. What they do show, given that it can safely be assumed that such figures were more likely to underestimate than exaggerate the numbers of wage-earning women, is that waged labour was a prominent part of the lives of many Italian women. Beyond such generalizations they reveal little. Only a closer look at these broad areas of female employment can shed light on women's contribution to and experience of work in this period.

Some studies have been made of female employment under Fascism but these are still surprisingly few in number. Official pronouncements, descriptions of laws, and analyses of census figures, important as they may be, can give only a limited picture. Many questions remain to be answered about women's experience of work in this period, whether it be about gender divisions on local labour markets and in the workplace, female attitudes to waged labour in a period of hysterical rhetoric directed against it, or the manner in which home and work lives interacted for those women who did work outside the home.

MAGNETI MARELLI

This book concentrates on a case study of a group of northern working-class women all employed outside the home. It focuses on the experience of women in one specific, albeit somewhat untypical, environment, a light engineering firm located in Sesto San Giovanni, already one of the most important industrial towns in Italy. Founded in 1919, by the Second World War Magneti Marelli had become one of Italy's industrial giants and arguably the most modern firm of its day. About half its workforce was female and its management methods rivalled those of many contemporary American firms. The history of the meteoric rise of this firm typifies the 'modernizing' aspects of the Fascist period in Italy. Unquestionably many economic sectors were severely hampered by the archaic and irrational policies of the government and for them the inter-war period was one of struggle against stagnation or even decline.[31] Nevertheless some firms managed to thrive. The example of Magneti Marelli

[31] There has been considerable historical debate about whether Fascist economic policy was modernizing or essentially politically motivated with a

shows how certain firms, especially if politically well connected and making at least some defence-related products, were able to expand and modernize.

The very modernity of this firm makes it significant in the history of Italian industry. It was a pioneer in many respects. Here, more than in any other firm of its day, Fordism and Taylorism came to Italy. Experiments in new production methods were wide-ranging, including not just conveyor belts and timekeeping but also a far more sophisticated attitude to motivating workers than some of the crasser, more heavy-handed approaches which prevailed in many firms in this period. Here, inspired by foreign ideas of 'human relations', the carrot as well as the stick was employed, aimed at moulding the 'human factor' of production in 'the most modern and scientific manner'.

Politics inevitably permeates this account of the history of Magneti Marelli and its work-force. It provides part of the explanation for the firm's phenomenal growth in the shape of government contracts. The political situation undoubtedly facilitated the introduction of aspects of scientific management which were resisted by collective bargaining in other more democratic countries. Politics also intruded on the daily life of the firm through its mass leisure organization the Dopolavoro, in its welfare facilities such as its nursery, and occasionally in full-blown pomp and ceremony when Fascist dignitaries visited this model firm in person.

None the less, politics in Fascist Italy frequently failed to have the great impact on daily life that 'Big Brother' theorists of totalitarianism might have us believe, and the fortunes of both firm and workers were shaped by a series of factors of which political ideology and policy was only one. The relationship between the Italian working class and Fascist politics is a complex one, as the intense historiographical debates about the extent of their consent or resistance to the regime have shown.

negative effect on economic growth and development. The literature on Fascist economic policy is too vast to list here but two useful articles in English which consider aspects of this debate are P. Corner, 'Fascist Agrarian Policy and the Italian Economy in the Interwar Years', in J. A. Davis (ed.), *Gramsci and Italy's Passive Revolution* (London, 1979); J. Cohen, 'Was Italian Fascism a Developmental Dictatorship?', *Economic History Review* (1988).

The case study of Magneti sheds light on such debates and in particular on the question of the formation of political consciousness amongst women, a group which is all too often neglected in histories of the 'making of the Italian working class'.

With regard to the management, their general political allegiance was undeniable, but the firm they ran failed to match up to many ideals espoused in Fascist rhetoric. It was a thoroughly modern industrial firm at a time when official ideology preached the virtues of the rural idyll. Its inspiration was clearly American in a period when things Italian were most praised. It was also an employer of large numbers of those who in Fascist ideology were better fitted to be 'angels of the hearth'.

This is as much the history of a particularly innovative and pioneering firm as it is a case study of female employment. It focuses on gender divisions and on the experience of the women in the work-force, but it is also a study of the introduction of scientific management and the ways in which this was moulded by the particular political and economic climate of the time. In many ways this study is new, for although it reflects the increasing number of historical studies of individual firms that have appeared in the last two decades (and continue to be published), it is one of the first to look at the complex history of an Italian firm from a feminist point of view.

1

Flashes and Glows: The Growth of Magneti Marelli

TODAY Sesto San Giovanni is a suburb, engulfed by the urban sprawl surrounding the great commercial city of Milan. Even so, it remains very much a town in its own right, with its own distinctive history and culture. For Sesto, since the early twentieth century, has been the Italian Sheffield, the 'city of the factories', with the largest industrial concentration in the Milanese area, dominated by steelworks and big engineering plants. It has long been known for its left-wing politics. A strong labour movement developed in the early decades of this century and, despite eventual defeat at the hands of the Fascists, a thin strand of this political tradition survived the grey years of Fascism, and the Sesto working class revived to earn their town the name of 'Stalingrado' during the Resistance. Even today, despite social change, economic problems, and even closure for some of the older plants, the reputation of 'Red Sesto' persists.

Large-scale industry appeared early in Sesto. Until the late nineteenth century, it was just a small town a few miles north of Milan, indistinguishable from numerous others in the northern part of the Po valley plain which largely depended on agriculture, textiles, and a few other small industries. But with the opening of the San Gottardo pass in 1882 Sesto's position on the Milan–Monza railway line became significant. This led to the development of some small engineering enterprises in the town. Other factors, such as the relative cheapness of land, the closeness of Milan,[1] and the fact that Sesto lay in a well-drained area which was far healthier than the swampy rice-growing lands further south, combined to encourage the rapid transformation of this sleepy, rural town into one of Italy's most important industrial

[1] It was, in fact, precisely in this period, at the end of the 19th cent., that Milan really began to emerge as the economic capital of Italy. See E. Dalmasso, 'Milan, capitale économique de l'Italie', thèse de doctorat (Aix-Marseille, 1970).

centres. At first most of the new industries, many founded by Milanese industrialists who had decided to transfer their production there, were small. But between 1903 and 1911 a number of larger firms such as Breda began production and established Sesto's character as the most industrialized zone of the Milanese area, with a number of large heavy engineering factories and steelworks.[2]

Most of the jobs in these plants were reserved for men (many of these being skilled workers), with only a few small textile firms and a handful of other industries such as the Pirelli rubber factory and a box-making firm offering employment to women. In this situation, therefore, the foundation of Magneti Marelli in 1919, a light engineering firm with mass production methods and many women in its work-force, represented a new direction for Sesto's development.

MAGNETI MARELLI

Given its innovative character, it is perhaps a little surprising that the new firm was not a product of foreign ideas and capital or the subsidiary company of some other, technologically advanced firm. Half the capital was indeed provided by the Turin car firm Fiat,[3] but the real 'parent' of the new company was one of Sesto's older traditional heavy engineering companies,

[2] In 1911 the first industrial census revealed Sesto as having 37 engineering and metallurgical firms with a work-force of 5,349, compared with only 8 textile firms employing a mere 813. This same 1911 census also demonstrates to what extent Sesto was becoming a centre for large industry. In Milan in this same period there were 1,946 engineering and metallurgical firms but they only employed an average of 22.2 workers each, compared with the 37 Sesto firms who averaged 141.9 employees each. See G. De Vecchi, 'L'industrializzazione della fascia nord Milano e la localizzazione delle industrie a Sesto San Giovanni', in G. Petrillo (ed.), *La città delle fabbriche* (Milan, 1981), 43–8. On the development of the engineering and metallurgical industries in Sesto see also L. Bosisio, 'Nascita e sviluppo del settore meccanico e siderurgico', ibid. 56–64; P. Lincoln Cadiolo, *Sesto San Giovanni dalle origini ad oggi* (Milan, 1964); G. Rulfi, 'Un centro industriale lombardo: Sesto San Giovanni', *Rivista geografica italiana* (1955); D. Bell, *Sesto San Giovanni: Workers, Culture and Politics in an Italian Town, 1880–1922* (New Brunswick, NJ, 1986), ch. 1.

[3] Fiat was in a position to provide this capital because of large profits made during the First World War. See V. Castronovo, *Giovanni Agnelli: la Fiat dal 1899 al 1945* (Turin, 1977), 135.

Ercole Marelli & Co. Fiat's role was limited to financial support and to providing a market for many of the firm's goods, whereas it was Ercole Marelli & Co. which actually set up a shop for production and provided the necessary technology and most of the new firm's management.

Ercole Marelli (who was head of the firm bearing his own name until his death in 1922) was a self-made man. The son of an artisan, he began his working life as an engineering apprentice. Once trained he was able eventually to set up a small workshop. This grew and by 1905 he owned his own factory in Sesto. With the advent of the First World War, like many other engineering companies, his firm prospered with the increased demand for armaments. In the pre-war period he had been experimenting with magnetos[4] and distributors, so he was

Table 1.1. Rise in production (no. of products produced each year: 1919–36)

Year	No. of products	Year	No. of products
1919	4,000	1928	67,000
1920	6,000	1929	77,000
1921	13,000	1930	104,200
1922	22,800	1931	130,000
1923	36,500	1932	134,600
1924	53,100	1933	198,200
1925	90,600	1934	212,100
1926	93,700	1935	298,000
1927	83,000	1936	434,200

These figures come from 'Vent'anni di lavoro nella Fabbrica Italiana Magneti Marelli', *SB* 1 (1937), 2–72. They refer to the number of 'units' produced. It must, therefore, be seen as only a very rough indicator of growth, as a large expensive product such as a radio transmitter counts as no more than a small one like a windscreen wiper. If this is borne in mind, the growth in output is greater than appears from the table, as the overall tendency was towards producing more complex products in the later period. It must be noted, however, that these figures are published by the firm itself and there is no way of checking their validity.

[4] Magnetos (from which Magneti Marelli derived its name) are electric generators using permanent magnets, used in the ignition of internal combustion engines.

Table 1.2. Average size of Magneti work-force

Year	Work-force	Year	Work-force
1919	200	1929	1,650
1920	400	1930	1,700
1921	850	1931	1,800
1922	710	1932	2,100
1923	880	1933	2,810
1924	1,115	1934	3,750
1925	1,360	1935	4,950
1926	2,070	1936	6,830
1927	1,700	1937	7,000
1928	1,580	1938	7,000

The figures for 1919 to 1936 come from 'Vent'anni di lavoro nella Fabbrica Italiana Magneti Marelli', *SB* 1 (1937), 2–72. Although not specified, it is almost certain that these figures include workers, clerks, and management. They represent the total work-forces of all the plants and firms within the Magneti group. It is unclear from the source exactly whether these figures represent the work-force on one certain day of each year or whether they are an average for the whole year. The figures for 1937 and 1938 are from 'Assemblea generale ordinaria della Fabbrica Italiana Magneti Marelli', *SB* 3 (1938), 45, and 'Assemblea generale ordinaria della Fabbrica Italiana Magneti Marelli', *SB* 3 (1939), 43.

in a good position to exploit the war situation when the supply of magnetos, hitherto imported from Germany, was cut off. Magnetos were essential components for the engines of aircraft, which were being used militarily for the first time in this war. In 1917 a special magneto shop was set up. By 1918 production was transferred to a separate new plant and trials were under way of magnetos for car engines. In October 1919 the new factory was established as a separate limited company under the direction of Bruno Antonio Quintavalle. The initial capital was of 7 million lire, the work-force numbered about 200, and in the first year production was of a mere 4,000 magnetos.

From these small beginnings the firm's output rapidly rose. By 1925 the annual production was over twenty times the output of the first year (see Table 1.1). The work-force increased steadily (see Table 1.2) and the range of products was considerably extended to include numerous electrical parts for cars, aeroplanes, and trains, as well as radios and military equipment.

As production expanded, the original small building, adjacent to the Ercole Marelli works, was gradually enlarged and 1929 saw the opening of a second plant, the 'B', where a (largely male) work-force of about 150 produced accumulator batteries. During the 1930s other plants were added, such as the 'D' (destroyed in the Second World War) in nearby Crescenzago which made insulators for spark plugs and radios, and the 'C' in Sesto for metallurgical work. In 1939 construction began in Crescenzago of the new 'N' plant, destined to become the firm's major production unit. Throughout the Second World War production was gradually transferred here from the A site.

In the late 1920s Magneti became the leader of a group of companies with the creation of a number of subsidiaries which mostly manufactured semi-finished parts required by the main factory. By the outbreak of the Second World War the Magneti group consisted of no less than sixteen different plants scattered throughout Italy and two sales subsidiaries.[5] Although still of modest dimensions compared with giants like Fiat, the Magneti group was, by this time, of substantial size and the unquestioned leader in its field on the Italian market.

EARLY DAYS: THE 1920s

With the ending of the war the new firm was faced immediately with the problem of foreign competition, and it was right from

[5] These were: Magneti Marelli, A, B, C, D, and E plants. (The E plant, founded 1936, was in Apuania.) SAVEP (Sezione Autonoma Vetrerie Pavesi) in Pavia. Produced glass products. SAMAS (Sezione Autonoma Macchine ed Attrezzature Speciali) at Caravaggio (Lombardy). Produced specialized machine tools for the other plants. Destroyed in the Second World War. FIVRE (Fabbrica Italiana Valvole Radio Elettriche), three plants in Pavia, Florence, and Milan. Produced radio valves. SAVAM (Società Anonima Vetraria Affini Milano). Founded 1936. Made glass products mostly for FIVRE. STAR (Società Toscana Apparecchi Radiofonici), one factory in Livorno. Made radio equipment. SAEN (Società Anonima Elettrovetre Nazionale), one factory in Agrate (near Milan). ICS (Industria Composizioni Stampate), two factories, one in Canonica d'Adda and the other in Bregnano (both near Milan). Main products: insulators for magnetos and distributors, casings, and styled radio cabinets. SAMIMA (Società Anonima 'Mica' Impianti Minerari in AOI), one plant in Archico (in Italian colonies in Africa). Mining.

Of these firms Magneti was by far the largest, with a capital of L. 100,000,000 in 1940 compared with FIVRE's L. 10,000,000. All the others were even smaller than this. The Magneti group also contained the sales subsidiaries Radiomarelli and MABO (a joint sales company with the German firm Bosch).

this early period that production began to be 'rationalized' on American lines in order to make its products competitive with modern foreign firms like Bosch. Production concentrated initially on car parts such as magnetos and distributors, but gradually the range of products increased and by the late 1920s the firm was taking advantage of the expanding market for aircraft engine parts. Reflecting the unprecedented growth of the Italian economy as a whole in the early and mid-1920s, this was a period of rapid expansion for the firm.

But from 1927 this growth began to waver. This was partly due to the increase in foreign competition with the recovery of the German economy, but even more important was Mussolini's stubborn determination to revalue the lira to the 1922 sterling parity of ninety (the so-called *Quota Novanta* campaign). This politically motivated policy caused havoc for many sectors of Italian industry including the car industry, which inevitably affected the demand for many of Magneti's products.[6] The crisis was ridden out largely by product diversification and an export drive. The work-force bore much of the brunt of this, as cost cutting was facilitated by the downward revision of wages in the collective contracts signed by the Fascist unions. Thus the absence of effective collective bargaining must be seen as crucial to the survival of the firm. Some of the workers did not even get the new lower pay rates, since many were laid off, sometimes for extensive periods.[7]

THE 1930s

After the trials of the *Quota Novanta* period, Magneti seems to have survived the economic problems caused by the world economic slump relatively well. Output rose throughout the entire recession. The continuing profitability and expansion were achieved in three main ways. Firstly, the effects of the

[6] Exports of cars fell from by a third between 1926 and 1929. Fiat suffered less from this than smaller firms who found it hard to compete, so that by 1928 Fiat accounted for over 80% of Italian car production. See Castronovo, *Giovanni Agnelli*, 346.

[7] In one case 1,200 Magneti workers were laid off from 24 Dec. 1928 to 7 July 1929. See L. Longaretti, 'Il fascio di Sesto San Giovanni: storia delle camicie nere in un centro industriale 1919–29', test di laurea (Milan, 1983–4), 185.

depression were buffered by a series of state orders, in particular for accumulator batteries. Secondly, the sheer range of products made by the firm and its ability to adapt to circumstances, bringing out new products as others failed to sell, helped considerably. Finally, stringent cost cutting was crucial in a period of falling prices. Product designs were simplified and the work process streamlined. Again, much of the burden of this was carried by the work-force, and lay-offs and sackings of the least profitable workers continued.

Another factor significant in the survival of the firm in the early 1930s was the rapid expansion of radio production. In the inter-war period radio was a huge new growth industry in Western Europe[8] and in Italy too the market for radios grew considerably, although its development lagged far behind that of other industrialized nations.[9] During the 1920s this small Italian market was dominated by foreign manufacturers and it was difficult to use tariffs to protect the nascent Italian industry because it was dependent on imported parts. Italian production was still confined to small artisanal workshops. In this context, as Antonio Papa argues, 'The birth of Radiomarelli . . . represented an exceptional event in the history of the Italian radio industry. This was the first radio firm in Italy to be linked with a large engineering firm.'[10]

At first the new company had difficulty competing with foreign manufacturers, especially as it was precisely in this period that technological advances were enabling these foreign firms to cut production costs and prices drastically. But in

[8] As David Landes has argued, this growth was due both to rapid technological improvements in this period and to the fact that demand rose quickly for a product which, although a luxury in some respects, was quickly able to develop a mass market in countries such as the UK. D. Landes, *The Unbound Prometheus* (Cambridge, 1969), 431.

[9] In 1927 e.g. there were only 26,865 radio licence holders in Italy compared with 1,230,000 in Britain. The actual number of radio owners was certainly considerably greater than (perhaps even double) the official statistics. A. Papa, *La storia politica della radio in Italia*, i, *Dalle origini agli anni della crisi economica: 1924–1934* (Naples, 1978), 26. In 1930 only 0.43% of the Italian population owned a radio licence. By 1942 this figure had risen to 4.1%. The growth was, therefore, considerable. However, to put these figures into perspective, it is sufficient to note that in 1939, when the number of Italian licence holders reached the one million mark, the figures for the UK were 9 million and for Germany 13 million. See ibid. ii, *Dalla guerra d'Etiopia al crollo del fascismo: 1935–43* (Naples, 1978), 35.

[10] Ibid. i. 94.

September 1931 the introduction of a limited amount of tariff protection enabled Magneti to expand radio production and reduce its prices through economies of scale. The domestic market was further widened as prices fell, particularly with the launching of the cheap Alauda model in 1933. Yet even this was expensive compared with earnings, and prices never fell enough to really open up the market to the working classes.

With the foundation of FIVRE (Fabbrica Italiana Valvole Radio Elettriche, a valve-making subsidiary) the thorny problem of imported foreign parts was partially resolved, but Radiomarelli still could not really survive alone in the competitive world of radio production and in 1934 an agreement had to be signed with the giant RCA (Radio Corporation of America). The shadow of foreign competition remained, even during the autarky campaign, since, in contrast to many other products, high tariff barriers were not erected and large numbers of foreign radios continued to be imported. Despite these problems Radiomarelli quickly grew to dominate Italian radio production[11] and its output and sales rose rapidly.

REARMAMENT AND AUTARKY

On 3 October 1935 Italian troops invaded Ethiopia, thereby incurring the wrath of the League of Nations, which imposed economic sanctions. Far from being cowed, the Fascist government took advantage of the situation to launch a huge patriotic propaganda drive. Italy was to become self-sufficient economically through the so-called autarky campaign. Accompanied by a fanfare of endless propaganda, imports of consumer goods were restricted and the import of raw materials controlled. Sectors important to the war economy (such as chemicals) were expanded. Great emphasis was placed on the need to discover Italian sources of raw materials and to invent new 'autarkic' products to replace foreign goods.[12]

[11] The firm claimed to be producing 80% of all Italian radio production by 1932, only two years after its foundation (ibid. i. 96).

[12] On the autarky policy see, for example, G. M. Rey, 'Una sintesi dell' economia italiana durante il fascismo', in G. Toniolo (ed.), *L'economia italiana 1861–1940* (Rome–Bari, 1978), 299–311.

For Magneti the sanctions created numerous difficulties in obtaining necessary raw materials. But on the whole the firm did very well out of the war. Rearmament brought highly profitable state orders for its products such as high-altitude military aircraft magnetos. Even the self-sufficiency drive was advantageous in certain respects since it gave Magneti a virtual monopoly on the Italian market for many of its products. In the 'Buy Italian!' campaign firms were directed to purchase Italian goods rather than foreign imports, and for certain products, such as radio valves, there were no rival domestic manufacturers. In this situation, the problems of the early 1930s disappeared and output rose again. The improvement was so dramatic that between 1934 and 1936 production doubled.

With the ending of the Ethiopian War the firm experienced some problems, but these were buffered by the continuation of the autarky campaign, although profitability was maintained once again only by a certain amount of cost cutting. With most foreign markets closed the firm was obliged to resort to rather desperate measures such as attempting to export to new areas such as Chile, Siam, and the Italian colonies in Africa. In spite of this, a certain amount of growth continued.

WORLD WAR AND THE FALL OF FASCISM

These problems proved to be transitory, since rearmament for the Second World War brought in new orders. Special machine tools for armaments production were imported from Germany and production continued to rise. The major Magneti plants were declared auxiliary factories with the work-force classified as 'mobilized civilians' subject to military discipline, and treated as deserters if they left their jobs without authorization. The management too were placed under military supervision and could not sack or transfer personnel without prior permission.[13]

Magneti was important in war production as the sole domestic manufacturer of many car and aircraft components and also the major producer of radio transmitters and receivers.

[13] 'Dichiarazione di ausiliarità—FIMM', ACS, OI, b. 22, f. 6, 'Commesse belliche'.

When the German army arrived in September 1943, Magneti was declared essential to the war effort.[14] The invaders strove to maximize output, but this was largely in vain, since, as both Quintavalle's attempt to clear his name through his lawyer after the fall of Fascism and the testimonies of ex-workers reveal, workers and managers slowed or even sabotaged the work to keep output for the German war effort down.

THE REASONS FOR GROWTH

Throughout the inter-war period, even during most of the bleak years of the depression, this firm experienced a quite remarkable expansion. In 1945 the antifascist Comitato di Liberazione Nazionale (CLN) report on the firm expressed the following opinion on the reasons for this extraordinarily fast growth: 'Apart from the work of Quintavalle, the management, and the workers, the development of Magneti Marelli is due to the virtual absence of competition for the firm's products.'[15] Yet clearly the firm's expansion cannot be ascribed simply to these causes alone. It is true that the firm enjoyed a virtual monopoly on the domestic market for some of its output, but other factors too must be taken into consideration, such as government orders, the links with Fiat, rearmament, and the particular organization of the work process.

One important factor was undoubtedly the close links with Fiat, which was already indisputably the largest and most

[14] Firms of the FIMM group were of interest to the invading Germans, who dismantled the Florence FIVRE and STAR plants (both Magneti subsidiaries). The Magneti management was only able to prevent their removal to Germany by promising that these plants would immediately start production for the German war effort if they were reassembled at Cantu in the north of Italy. According to the lawyer of the Managing Director, they were never brought into production. (R. Garbagni, 'Promemoria in difesa di Bruno Antonio Quintavalle', ACLNAI, b. 58, f. 4, p. 7.)

[15] Lengthy investigations were made by the CLN (*Comitato di Liberazione Nazionale*—National Liberation Committee) of the firm to try to establish the 'guilt' or 'innocence' of the directors as collaborators with the Fascist regime. This extract is taken from the report on Bruno Antonio Quintavalle, compiled from testimonies of various employees of the factory and people who had known him. 'Rapporto sul Conte Bruno Quintavalle—Amministratore delegato della Magneti Marelli', ACLNAI, b. 43, f. 3.

important Italian car manufacturer in this period. The domestic car market was restricted, as road building was minimal and per capita income low. Italy was, as yet, a far cry from the USA where Ford's Model T had made car ownership possible for vast new sections of the population. But in spite of the problems, production of motor vehicles was expanding generally in inter-war Italy, with at times as much as half the output destined for export.[16] Apart from Fiat, Magneti also manufactured parts for numerous other smaller Italian car producers such as Lancia, Itala, and Bianchi and also for export.

Magneti Marelli did not, however, sell to these firms alone, and undoubtedly the firm's commercial and marketing strategy was also important. Advertising was fairly imaginative, making use of magazines and newspapers, sponsored sports events, the factory's own newspaper *Sprazzi e Bagliori* (Flashes and Glows),[17] international salons, and the organization of promotional events such as Radiomarelli concerts. Already in the 1920s the firm seems to have had a world-wide sales network, probably linked to the commercial outlets of Ercole Marelli & Co., and even in the difficult period of the 1930s, when world trade was greatly slowed down by protectionism on all sides, Magneti claimed to be continuing to export. An extensive distribution network was

[16] The Italian car industry was the fourth in Europe for production and the second in exports; 50% of car production was exported, due to the restrictedness of the Italian market. Although in 1931 Italian car manufacturers still only produced 1% of the total world car output this was, nevertheless, a sector in expansion.

[17] The factory newspaper was produced from 1924 until 1943. Copies of almost every issue still exist and they constitute an invaluable source of information for many aspects of life in the firm. The magazine was both part of the firm's consensus strategy, aimed at creating a kind of family atmosphere on the shop-floor, and also an element of its marketing campaign as many of those who read the paper were mechanics in garages and repair workshops all over Italy. For this reason the paper contained a varied mixture of news about the firm and its employees, technical articles, political propaganda, etc. It was very professionally produced and contained numerous illustrations and photographs. In the 1920s this magazine was edited by a member of the managing director's family, Noel Quintavalle, who seems to have been artistically inclined, producing cartoons and illustrations for the magazine as well as artwork for numerous Magneti advertisements, under the pseudonym 'Noelqui'. At first, when the firm was small, the newspaper was very simple, full of personal references and even comic cartoons of the management, but from 1929 it became an altogether more polished affair with a much larger format and looking more like a glossy magazine than the previous small news-sheet.

established from 1929 all over Italy of 'authorized sales outlets' and 'authorized repair shops' as part of the attempt to present an image of modernity and reliability. According to the factory newspaper, this commercial distribution system was crucial to the firm's drive for the 'Americanization' of its production.[18]

International agreements, a common feature in this sector where firms needed to keep abreast of constant technological developments, were signed with a number of foreign firms. Probably the most important were the links with RCA and with Bosch.[19] The licence agreement with RCA for the production of radio parts was signed in 1934[20] and this was followed a year later by a market sharing agreement with the German light engineering company Bosch in car and motor-bike parts. The marketing, sales back-up, and repair services of the two firms were united with the foundation of a joint subsidiary company, Mabo (Società Anonima per la vendita dei prodotti Magneti Marelli e Bosch).

Another important factor in the firm's growth was government orders, probably facilitated by the politically influential position of Benni, the firm's president. As early as 1930 the firm had become the sole supplier to the Air Force for all of its products. Other contracts included government purchases of accumulator batteries for railways and submarine batteries for the Navy. With the advent of war in the late 1930s this factor became crucial.

The firm's growth, therefore, can be ascribed to a number of causes. The rising demand for radios, the links with Fiat,

[18] 'Le nostre filiali', *SB* 9 (1929), 35. Special 2-week training courses were organized in the factory itself to train the personnel of these garages and workshops on how to fit and repair Magneti products. This ensured a good repair and back-up service for sales.

[19] Agreements were signed with numerous other firms including the American General Electric, Westinghouse, Bendix, Mallory, Lockheed, and Hazeltine and the British Chloride Storage Batteries, Alladin, ML (Lucas), Hydraulic Brake Company, and DP.

[20] RCA granted to FIVRE a non-exclusive, indivisible, and non-transferable licence to manufacture radio parts patented by RCA in Italy, or parts for which RCA had bought the Italian patents. In return RCA could manufacture from FIVRE designs patented in the US. RCA also undertook to provide detailed designs and technical advice. See 'Agreement between Radio Corporation of America and Fabbrica Italiana Valvole Radio Elettriche and Fabbrica Italiana Magneti Marelli', 18 July 1934, ACLNAI, b. 43, f. 2.

rearmament, and a number of important state orders all meant that it was in the fairly unusual situation of manufacturing for a relatively steady and expanding market. Equally significant were the marketing strategy, unusually advanced for this period, and the government autarky policy which gave it a monopoly for many of its products. The existence of the Fascist government was crucial to the firm's success in other ways, for it made strikes virtually impossible and created an atmosphere in which the management was able to experiment freely with new production ideas. Thus it was that the firm was able to introduce 'American' management methods without the pitched opposition of unions which had often dogged similar initiatives abroad.[21] It was, in the end, this willingness to organize along 'modern' lines which enabled Magneti Marelli to gain control of the Italian market so that export orders were won and the competition of other firms, sometimes little more than artisan workshops in fields such as radio production, simply melted away.

It is implicit in this account that, despite the contradictions of Fascist ideology, the management had much to gain from making its support for the government clear. On balance, despite some problems such as in the *Quota Novanta* period, it is evident that this firm benefited from government policies. All this makes nonsense of the claim, put forward in 1945 by the managing director's lawyer in an attempt to clear his client's name, that

> it is very clear that such an open and practical attitude to management meant that Quintavalle was bound to openly oppose the ideology and practice of Fascism which called for, as is well known, a narrow-minded policy of autarky and an irritating, stupid emphasis on nationalism.[22]

THE MANAGEMENT

The first president of Magneti Marelli was Antonio Stefano Benni, 'one of the most influential industrialists in the Fascist

[21] See e.g. on the introduction of Fordist ideas in Bosch in the pre-war period, H. Homberg, 'Anfänge des Taylorsystems in Deutschland vor dem Ersten Weltkrieg: Eine Problemskizze unter besonderer Berücksichtigung der Arbeitskämpfe bei Bosch 1913', *Geschichte und Gesellschaft*, 2 (1978).

[22] Garbagni, 'Promemoria in difesa di Bruno Antonio Quintavalle', 3.

period'.[23] Born in Cuneo in 1880, he began his career as a clerk in Ercole Marelli's small engineering business.[24] As this firm grew, so Benni's fortunes prospered, rising to become Marelli's second-in-command. Put in charge of sales and exports he got the chance to travel widely. On the death of Ercole Marelli, he became managing director of the original firm. Around this time he also began to accumulate numerous posts on the boards of other firms.[25]

Paralleling his growing importance in industrial circles was his interest in right-wing politics. In 1917 he got involved in an interventionist group, the *Comitato milanese di azione interna*, one of whose members was none other than Benito Mussolini. Initially, however, he joined not the Fascists but the Nationalist Party. A founder member of the influential employers' organization *Confindustria*, in 1919 he was part of the industrial delegation which attempted to negotiate with the engineering workers shortly before the 'occupation of the factories'. Elected a Liberal-Nationalist deputy in May 1921 and president of the *Confindustria* in 1922, he supported the Fascists, believing with many of the other right-wing deputies that Mussolini could be tamed and controlled for their own ends. When this proved the death blow to liberal democracy he adhered to the new government, joined the Fascist Party in January 1926 after the Vidoni Palace Pact, and later became a member of the Fascist Grand Council.

As head of the *Confindustria* he created enemies among more 'left-wing' Fascists by his opposition to the Corporate State, and as a result was eventually removed from this position in 1933.[26] This was not, however, the end of his political career, as from May 1934 to January 1935 he chaired the government's Budget Committee and then became Minister of Communications until

[23] P. V. Cannistraro, 'Benni Antonio Stefano', in Cannistraro (ed.), *Historical Dictionary of Fascist Italy* (Westport, Conn., 1982), 67.

[24] Unsigned document, Milan, 5 Sept. 1924, ASM, cartello 57, sf. 1487 'Benni'.

[25] Promemoria 'Benni', ibid.

[26] It has been argued that in his position as president of the *Confindustria* Benni was overshadowed by the much more cultured and educated Olivetti. But others maintain that the two complemented each other, since although Benni lacked education and culture he was practical and unprejudiced. See P. Melograni, 'Benni Antonio Stefano', in *Dizionario Biografico degli Italiani*, viii (Rome, 1966), 559.

October 1939, leaving some of his business commitments to take up this post, including his presidency of the board of directors of Magneti Marelli. In 1939, however, he lost this ministerial position and his deputy's seat and thereafter largely abandoned politics to devote himself to his business career. Despite his links with Fascism he was arrested in 1944 by the Fascist police because of his refusal to give his support to the Republic of Salò. But, fearing for his life from both Fascists and antifascists, he managed to flee to Switzerland, where he eventually died on 28 December 1945.[27]

Benni, therefore, like many other key industrialists of his time, was not a 'Fascist of the first hour' but, solidly committed to right-wing politics, he was quick to recognize the potential of the Fascist regime.[28] His attitude to Fascism can best be termed opportunistic. Whilst retaining a certain critical distance from some of Mussolini's policies, such as the *Quota Novanta* campaign (and also the idea of the Corporate State, which interested him only insomuch as it furthered the needs of the employer),[29] he was a loyal supporter of the government, recognizing the power that it gave him and others like him in their own firms. A self-made and highly ambitious man, Benni's

[27] See Cannistraro, 'Benni Antonio Stefano'. On Benni's career see also *SB* 2 (1935), 2–3, and E. Pozzoni, *Commemorazione di Antonio Stefano Benni* (Milan, 1959). (This is a copy of a speech given by Pozzoni in his capacity as Presidente dei Cavalieri del Lavoro on 23 Apr. 1959.) See also Melograni, 'Benni Antonio Stefano', and 'Corporazione della metallurgica e della meccanica' (1934), ASM, cartello 1062, f. 4. For an admiring and romanticized account of Benni's life by Cesare Rossi, ex-political Secretary of the Milanese Fascio, see 'Antonio Stefano Benni: un capo venuto dalla gavetta', in C. Rossi, *Personaggi di ieri e di oggi* (Milan, 1960), 151–3.

[28] On the attitude of the industrial leadership to Fascism see P. Melograni, *Gli industriali e Mussolini: rapporti tra Confindustria e fascismo. Dal 1919 al 1929* (Milan, 1980). Melograni's general argument is that, in the 1920s, industrialists supported Fascism as far as it suited them but with a certain reserve, preferring to dedicate themselves to longer term aims, which went beyond the regime. This somewhat distant and suspicious attitude described by Melograni for the 1920s was mostly replaced by a far greater support by the 1930s when it seemed as if the regime would last. See R. Sarti, *Fascism and the Industrial Leadership, 1919–40* (Berkeley, Calif., 1971).

[29] Another way in which Benni demonstrated his slightly critical attitude to the Fascist laws lies in the fact that he protected Gino Olivetti, who had been secretary general of the *Confindustria* whilst he was president, from the racist laws introduced in the late 1930s. (Melograni, 'Benni Antonio Stefano'.)

influence in the government undoubtedly played a part in the prosperous fortunes of the firm.

The other two most important personalities in the Magneti top management, the Quintavalle brothers, Bruno Antonio and Umberto, contrasted with the self-made Benni, for they had far more formal training and came from a totally different social milieu. Born in Sardinia, they were raised in Milan. Their father was a secondary school headmaster and their family aristocratic although not particularly rich.[30]

Umberto, the younger of the two, rose to prominence through the patronage of his brother. Even so, it is clear that in technical terms his was the crucial influence on the fortunes of the firm. The fact that he was a trained engineer seems to have been decisive in the 'rational' reorganization of production. When new technology was to be purchased or factories visited in the USA, it was Umberto who was sent on the trips. Unlike many industrialists who, suspicious of its populist rhetoric, had mostly come to support Fascism only after the March on Rome, Umberto was a fully fledged Fascist right from the beginning in 1919. He took an active part in the destruction of the Milanese left as the founder and commander of the 'Lupi' Fascist action squad. In addition to his leadership, he provided his gang of louts with transport for their various raids in the shape of a Magneti Marelli truck. In 1924 he joined the Sesto branch of the Partito Nazionale Fascista (PNF).[31]

His elder brother Bruno Antonio held the post of managing director from the foundation of the firm until long after the Second World War. He was trained in business studies at the Commercial Studies Institute in Milan from 1908 to 1911 worked in London. On his return to Italy he obtained a clerical position in the Ercole Marelli export department. But it was with the First World War that his career really took off, since it was then that, whilst recovering from a head wound after fighting as a cavalry officer, he met and eventually married Ercole Marelli's eldest daughter Paola who was working as a nurse.[32] This romantic

[30] 'Rapporto sul Conte Bruno Antonio Quintavalle—Amministratore delegato della Magneti Marelli', ACLNAI, b. 43, f. 3, p. 1.

[31] *Il popolo di Lombardia*, 18 Feb. 1929. See also 'Uomini della Magneti Marelli', *SB* 6 (1936), 3.

[32] By this marriage Quintavalle also became a rich man, since his bride brought a dowry of 40,000,000 lire.

encounter proved the turning point in his career, since, no longer a mere clerk but the boss's son-in-law, he was promoted to become head of the new Magneto Shop in 1919, and when a separate company was formed later that year it was put under his command. On the death of Ercole Marelli in 1922 he became managing director of the new enterprise.

A rare testimony by one of the firms engineers furnishes a glimpse of Bruno Antonio's character:

> Quintavalle has a reserved and haughty nature. He is rather authoritarian and not very sociable. Because of his character in general he is not much liked by his employees. Quintavalle is an intelligent and active person and in spite of appearances he is good and honest.[33]

Perhaps not popular with the work-force, Bruno Antonio was well liked by the Fascist government, which heaped honours upon him, making him a *Grand'Ufficiale della Corona d'Italia* in 1936 and a count in 1938. Politically he was a classic example of a 'fellow traveller' of Fascism. According to one fairly sympathetic report (submitted to the firm's *Comitato di Liberazione* after the war) he was

> descended from a patriotic family. He always had liberal ideas. He joined the Fascist Party and received an honorary membership card from the Sesto San Giovanni *fascio*. Initially, at the rise of Fascism it seems that he sympathized with the new party. Later, however, he seems to have come to oppose it. He never held office in the party. He did hold positions in industrial and corporative organizations of the period but he was offered these because of his technical competence and not for political reasons.[34]

It is true that in the early 1920s he greeted the new party with enthusiasm, obtained a party card, and even took part in some 'squad' actions with his brother. In the final years of the regime, however, his attitude became more ambivalent. The assertion that he came to oppose Fascism refers to the fact that, although the autarky policy brought profits for Magneti, in the long term it meant problems for a firm with so many international links. For this reason, Quintavalle grew critical of the government's policies and became the target of a short-lived campaign in the

[33] Testimony of Ing. Bacchini, in 'Rapporto sul Conte Bruno Antonio Quintavalle'.

[34] 'Rapporto sul Conte Bruno Antonio Quintavalle'.

Fascist newspaper *Il Popolo d'Italia*, accusing him of being 'anti-autarkic' and 'dominated by foreign interests'.[35] Eventually in 1941 Bruno Antonio lost his PNF membership card, clearly largely due to this campaign but actually sparked off by an incident in which he refused to help an employee who was ill. On the letter requesting help Quintavalle angrily wrote: 'Who is this upstart?' A clerk took this letter to the mayor and Quintavalle was stripped of his party card, probably because the employee asking for help was a soldier.[36] He never got his card back.[37] In April 1944 he was arrested by the SS (a fact he was keen to stress after the Liberation) and briefly taken to San Vittore prison accused of helping the partisans, before being released due to lack of evidence.[38]

With the Insurrection and the Liberation Bruno Antonio fled fearing for his life. Once the fighting was over, the firm was temporarily placed in the charge of one of the firm's engineers appointed by the Liberation Committee, Leonardo Brasca,[39] who ran the firm for about eighteen months assisted by a management committee composed of representatives of the various resistance parties. But Quintavalle wanted to return and a long polemic ensued between the Liberation Committee of the

[35] Quoted in F. Guizzi, 'La Fabbrica Italiana Magneti Marelli', in G. Petrillo and A. Scalpelli (eds.), *La ricostruzione nella grande industria: strategia padronale e organismi di fabbrica nel Triangolo 1945–1948* (Bari, 1978), 261.

[36] 'Rapporto sul Conte Bruno Antonio Quintavalle'.

[37] According to the testimony in Quintavalle's defence written by his lawyer, it appears that for almost a year he had to leave the firm to go into hiding. However, I have found no trace of this story elsewhere.

[38] 'Promemoria', 1 Aug. 1945, ACLNAI, b. 43, f. 3, p. 3. According to this same source, Quintavalle did in fact help the partisans.

[39] Guizzi notes: 'Ingegnere Leonardo Brasca was the only one of the top management who had not compromised himself with the Fascist regime and had not associated himself, during his past career, with the intimate circle of the cronies of the managing director and his brother, Ingegnere Umberto Quintavalle . . . It was, therefore, considerations internal to the firm which led to the choice of a man who, although without particular political merit, at least was "untainted" by the top managerial group, whose main characteristics, under the influence of Quintavalle, were lack of moral values, corruption, and greed. Besides this, Brasca's own personality made him a good person to mediate and he was capable of negotiating expertly with the shareholders' (Guizzi, 'La Fabbrica Italiana Magneti Marelli', 245–6). Brasca's title was officially *Commissario Straordinario per l'Alta Italia della Soc. Magneti Marelli*. Brasca, who was a trained engineer, was head of the Commercial Department ('Curriculum vitae dell'Ing. Brasca Leonardo', ACLNAI, b. 58, f. 4).

Magneti subsidiary FIVRE, who asked for him to be brought back, and the Sesto Magneti Committee which wanted to be rid of him.

The accusations levelled against the Quintavalle brothers were many and serious. They were alleged to have amassed scandalously large profits under the Fascist regime and hidden some of this money abroad. They were also condemned for disseminating Fascist propaganda, including propaganda in favour of war, in the hope of increasing their profits. In general they were said to have been inflexible in their attitude to the workers' demands for better conditions during the war, and to have favoured the rise of a strata of nonentities to middle-management positions, preferring mere lackeys to persons who really knew their job. Other accusations were more specific. After the 'B' Plant had been bombed, for example, half the personnel were sacked and therefore exposed to the risk of being deported to work in Germany because they were unemployed. The managers were also said to have abused the powers they had as a 'mobilized' firm, preventing, for example, certain skilled workers from leaving the payroll if they had been offered better jobs elsewhere. Umberto was singled out for his active membership of the Fascist Party and for the fact that he was said to have favoured promotions for workers or clerks willing to marry female employees who had been his lovers. Furthermore, the brothers were condemned because they had abandoned the firm at the moment of the 'Insurrection', which was seen as proving their guilt.[40]

The Regional Liberation Committee was called in to adjudicate, and after lengthy deliberations and endless reports they were finally forced to allow Bruno Antonio to return.[41] In theory the reasons were the following:

> . . . the accusations levelled against him are not covered by the Purge Laws.
>
> The Regional CLN has also verified that Quintavalle cannot be accused of collaborating either with the German invaders or with the Fascists since, on the contrary, they persecuted him for years . . .

[40] CLN FIMM Comunicazione no. 033, ACLNL, b. 4, f. 13.

[41] Undated document signed 'Il CLN Aziendale', and 'Il CLN Regionale', ACLNAI, b. 4, f. 13.

In reality, however, the situation was more complex. Whether they believed the Quintavalles guilty or not of this litany of crimes and misdemeanours became irrelevant, since the livelihoods of the FIVRE work-force were at stake. The managing director had to be brought back due to pressure from America in the shape of RCA, on which FIVRE had been technically dependent since 1934. RCA simply refused to work with the new management, 'tainted', as they saw it, by communism.[42]

To complete the picture of the most important members of the management, Michele Soliveri should be mentioned. Coming from a poor rural family, he moved to Milan in search of work. After a spell as a restaurant worker, in about 1907 he became a worker in the firm of Ercole Marelli. From these humble beginnings he rose gradually until by the 1920s he had become works manager of the 'A' Plant.[43] A keen Fascist, albeit one with a socialist past, he seems to have made many enemies and during the Liberation was killed by the Resistance.

A FASCIST FIRM?

Despite attempts to deny this, it is evident that Magneti was run by men who were closely implicated in the Fascist regime, and they in turn benefited from it both in the sense of winning government orders for their products and of enjoying a union-free atmosphere which gave them great room for manœuvre with the work-force. Without the threat of strikes cost cutting could be passed on to the workers, and management had a free hand to introduce new experimental production methods and pay systems. The management's political sympathies left their mark on many aspects of life in the factory. The factory newspaper was frequently crammed with propaganda articles glorifying Fascism, almost every possible occasion was seized

[42] This is but one small example of the concerted effort by the Americans to avert what they perceived as a threat of Communist revolution in post-war Italy. On the anticommunist campaign see D. Ellwood, 'From Re-education to the Selling of the Marshall Plan in Italy', in N. Pronay and K. Wilson (eds.), *The Political Re-education of Germany and her Allies after World War Two* (London, 1985).

[43] For a pompous and self-glorifying account of his own rise see M. Soliveri, 'Venti anni di lavoro: un periodo—un ricordo, 5 nov. 1907–5 nov. 1927', AISRMO, CLN, Carte Mantovani.

upon to make propaganda for the government, and frequently there were visits from foreign dignitaries and party officials. The factory shops themselves were hung with placards bearing Fascist slogans.

Even Bruno Antonio, who eventually fell into disfavour with the Fascist authorities, must be seen in this light. As Febo Guizzi quite rightly argues:

> The independence which he displayed and the fact that he did not always agree entirely with government policy, do not fundamentally contradict the privileged position he was guaranteed by Fascism, and in some ways they simply confirm the well-known fact that important industrialists tended to support the regime whilst retaining a certain amount of autonomy; in other words they gave their support to a government which upheld their class dominance, without going as far as to maintain that support to the end, especially in the precarious, doomed phase of the Republic of Salò.[44]

The management, therefore, despite moments when they attempted to distance themselves from the regime, were, in many respects, a Fascist management. As for the workers, their political views are harder to gauge. They worked in a factory marked by propaganda and Fascist rhetoric but they also lived in or near to 'Red Sesto' with its own left-wing political tradition. I shall return to the question of the workers' political attitudes in the final chapter, as this issue cannot be adequately broached without first looking more closely at the nature of the firm itself, those who worked in it, and management policies which aimed at influencing their opinions and attitudes.

[44] Guizzi, 'La Fabbrica Italiana Magneti Marelli', 262.

2

'Taylorismo' at Magneti I: The Organization of the Work Process

THE ideology of rationalization and efficiency inspired by American 'scientific management'[1] aroused a considerable amount of interest in Europe, although, in contrast to the United States where the commitment to technological efficiency had pervaded the whole culture,[2] in Europe the appeal of scientific management was more selective and the greatest enthusiasm was found in countries where representative government was in crisis and radical alternatives were being proposed. The appeal of the American vision in Europe did not, however, follow the traditional Left–Right divide. Taylor's and Ford's European followers ranged across the political spectrum from Lenin and Gramsci to Hitler.[3] Actual experiments with scientific management also began early in Europe, first in the British firm

[1] The question of scientific management has aroused considerable historical and sociological debate in the last two decades. Much of this has been inspired by Harry Braverman's study of the changes in work organization in 20th-century America, *Labor and Monopoly Capital: The Degradation of Work in the Twentieth Century* (New York, 1974). Although many of Braverman's ideas have been criticized or even discredited, his work remains an important one. For a wide discussion and good bibliography of the debate on Braverman's ideas see S. Wood (ed.), *The Degradation of Work? Skill, Deskilling and the Labour Process* (London, 1982).

[2] For an acute and well-argued analysis of how this was effected by corporate engineers see D. Noble, *America by Design: Science, Technology, and the Rise of Corporate Capitalism* (Oxford, 1977).

[3] C. S. Maier, 'Between Taylorism and Technocracy: European Ideologies and the Vision of Industrial Productivity in the 1920s', *Journal of Contemporary History*, 2 (1970). On the wide-ranging appeal of Fordism see also P. Ortoleva, 'Henry, Heinrich, Enrico Ford: il fordismo come luogo comune nella cultura europea degli anni venti', introduction to H. Ford, *La mia vita e la mia opera*, trans. S. Benco (Milan, 1980). Lenin was at first critical of Taylorism, but eventually some of Taylor's ideas were applied in the effort to raise productivity with the 'New Economic Policy', although this did cause a great deal of ideological debate in the Soviet Union. See R. Traub, 'Lenin and Taylor: the Fate of "Scientific Management" in the (Early) Soviet Union', *Telos*, 37 (1978).

Hopkinson's in 1905 and then in France, notably at Berliet and Renault in 1910. Stimulated especially by the First World War, which created large markets for certain goods and saw the influx into industry of a great deal of unskilled, often female, labour, scientific management spread, although with the ending of the war and the consequent contraction of markets, problems arose in some sectors. The restricted size of markets was perhaps the main reason for the fact that, compared with the USA, in spite of the amount of interest in technocracy and scientific management, Taylorism and Fordism[4] were not widely introduced into European industry in the interwar period. In fact, the only country to adopt American methods on a large scale was Germany, where they were introduced in many firms especially in the period following the First World War.[5]

Modern management methods also aroused a considerable amount of interest in interwar Italy. Gramsci, for example, devoted some of his prison notebook writings to scientific

[4] Taylorism and Fordism were not, of course, identical although they are frequently referred to together almost as if they were. There were many differences between the ideas of the two men. Taylor was first and foremost an engineer with a technical approach to the problems of production. In some ways he was an idealist compared with the pragmatic entrepreneur Ford, whose main aim was profit, and the Fordist model was more authoritarian and alienating for the worker. For an examination of the differences between the ideas of Taylor and Ford see A. Accornero, 'Dove cercare le origini del taylorismo e del fordismo', in *Il Mulino*, 241 (1975). On Taylor see F. B. Copley, *Frederick W. Taylor: Father of Scientific Management*, 2 vols. (New York, 1923); S. Haber, *Efficiency and Uplift: Scientific Management in the Progressive Era* (Chicago, 1964). See also their own writings, F. W. Taylor, *Scientific Management: Shop Management, the Principles of Scientific Management, Testimony before the Special House Committee* (Westport, Conn., 1972); H. Ford (in collaboration with S. Crowther), *My Life and My Work*, new edn. (London, 1924; first published 1922).

[5] On the spread of scientific management in Europe, see C. S. Maier, *Recasting Bourgeois Europe: Stabilization in France, Germany and Italy in the Decade after World War I* (Princeton, NJ, 1975); Maier, 'Between Taylorism and Technocracy'; A. Pichierri, 'Diffusion and Crisis of Scientific Management in European Industry', in S. Giner and M. Scotford Archer (eds.), *Contemporary Europe: Social Structures and Cultural Patterns* (London, 1978); P. Devinat, *Scientific Management in Europe*, ILO, Studies and Reports, Series B, 17 (Geneva, 1927). Germany was the European country which saw the most widespread introduction of American ideas. The German model of rationalized industry differed from some other countries in that there was a great concentration on the role of industrial integration to regulate output. See W. Meakin, *The New Industrial Revolution* (London, 1928); and also L. Burchardt, 'Technischer Fortschritt und sozialer Wandel: Das Beispiel der Taylorismus in Deutschland', in W. Treue (ed.), *Deutsche Technikgeschichte* (Göttingen, 1977).

management, which he saw as an important key to understanding the structural changes in Europe in this period. He heralded Fordism as having great revolutionary potential, although he was, at the same time, critical of it as a 'revolution from above'.[6] The Fascists also were attracted by the themes of technocracy and productivism from the early days before their rise to power.[7] What was so appealing about Taylor's work was that it offered more than merely a new method of raising output in a factory (Taylor was, after all, not the first to suggest rationalizing management methods): Taylorism was interpreted as meaning a new approach to industrial relations and, through its claim to be 'scientific' (and thereby supposedly unbiased), offering a vision of a new classless society.

Taylor argued that a society was possible in which classes co-operated and trade unions were unnecessary if worker and boss worked together in harmony aiming for the joint goal of higher productivity. His system, he maintained, would make trade unions obsolete[8] and his ideas represented nothing less than a 'technocratic utopia'[9] with great appeal for those who dreamed

[6] A. Gramsci, *Americanismo e fordismo* (Turin, 1975). See also Ortoleva, 'Henry, Heinrich, Enrico Ford' for a comparison of Gramsci's analysis of Fordism with that of other antifascist and Fascist theorists.

[7] On early Fascist attitudes to technocracy see M. Vaudagna, 'L'américanisme e le management scientifique dans l'Italie des années 1920', *Recherches*, 32–3 (1978).

[8] See, for Taylor's ideas on trade unions, Taylor, 'Shop Management', in *Scientific Management*, 182–95. For example: 'This state of affairs is far from satisfactory to either employers or men, and the writer believes the system of regulating the wages and conditions of employment of whole classes of men by conference and agreement between the leaders of unions and manufacturers to be vastly inferior, both in its moral effect on the men and on the material interests of both parties, to the plan of stimulating each workman's ambition by paying him according to his individual worth, and without limiting him to the rate of pay or work of the average of his class. The amount of work which a man should do in a day, what constitutes proper pay for this work, and the maximum number of hours per day which a man should work, together form the most important elements which are discussed between workmen and their employers. The writer has attempted to show that these matters can be much better determined by the expert time student than by either the union or a board of directors and he firmly believes that in the future scientific time study will establish standards which will be accepted as fair by both sides' (pp. 186–7).

[9] See G. Bonazzi, 'Il taylorismo tra strumento del capitale e utopia tecnocratica (II)', in *Economia del lavoro*, 2 (1972).

of a new model of class relationships to replace class conflict. As Charles Maier has pointed out:

Fascist, or more precisely national-fascist ideology, resembled Taylorism in a key particular. It promised a non-zero-sum world in which classes no longer prospered only at each other's expense, in contrast to the implications of the traditional spectrum of European ideologies which were all zero-sum or redistributive . . . Expanding productivity meant that no repartition of a fixed quantum of national wealth was required. Postulating a new social category of producers or, more narrowly, an elite of scientific managers who arbitrated conflict meant that the hostile confrontation between the traditional classes was superseded.[10]

Yet Taylor's ideas on unions differed from the Fascist corporatist vision in that he did recognize the beneficial effects that unions had had in the field of winning better conditions for the working classes and in that he saw the way to win class co-operation as through high wages, which was never a feature of the Italian form of scientific management.[11] Taylor's belief in the obsolescence of unions proved, however, to be crude and simplistic, since the introduction of his methods led to considerable industrial unrest.[12]

The role of the Fascists in creating an environment favourable to the introduction of Taylorism should not be underestimated. The destruction of free trade unions enabled employers to innovate in the manner they saw fit. This moulded not only the extent to which scientific management was introduced but also its particular form; in the union-free atmosphere of Fascist Italy it was easy for employers to introduce the American model selectively, ignoring innovations they considered too costly.

Thus the Fascist government contributed to the introduction of rationalization into Italian industry less by actual intervention to impose scientific management methods than by creating the

[10] Maier, 'Between Taylorism and Technocracy', 43.

[11] The introduction of scientific management in Italy was not accompanied by a high wage policy. The only incident where an important industrialist advocated a different strategy from this was when Giovanni Agnelli (managing director of Fiat), in an interview with United Press in 1932, called for a Keynesian solution to the slump, namely: 'Cut the working day and raise hourly wages proportionately.' See V. Castronovo, *Giovanni Agnelli: la Fiat dal 1899 al 1945* (Turin, 1977), 374–6.

[12] See H. G. J. Aitkin, *Taylorism at Watertown Arsenal* (Cambridge, Mass., 1960). See also Noble, *America by Design*, 269–72.

conditions in which it could be implemented. Nevertheless, in 1926 the Fascist government did become more directly involved in the diffusion of American management ideas with the foundation of ENIOS (Ente nazionale italiano per l'organizzazione scientifica del lavoro). In the 1920s ENIOS was very active in publishing information on scientific management techniques, especially through its journal *L'organizzazione scientifica del lavoro*. One of its main roles was to foster the spread of rationalization in small and medium-sized firms, thereby closing the gap between small backward industries and the more advanced large firms. This organization did not confine its interest in rationalization to industry, and campaigned for the spread of 'rational' methods in other spheres of life. Even housework was to be 'rationalized', championed by ENIOS's publication *Casa e lavoro* (Home and Work). In the 1930s, however, the role of ENIOS changed and it lost its modernizing and Taylorist character to become largely a propaganda agency for the government, concerned with pushing its warmongering autarkic policies.[13] With the advent of the Second World War there was a renewed urgent need to raise productivity, which led to a revival of ENIOS's earlier role, but because of the fall of Fascism, there was little time for this new policy to have much concrete effect.[14] Duccio Bigazzi has commented:

> As one fairly well-informed expert has noted, in practice methods were adapted and virtually 'reinvented' in each factory: because of this, factories using modern methods fairly similar to those in other European firms coexisted side by side with extremely backward firms and in the war economy these out-of-date firms inevitably dragged down the efficiency of the others.[15]

Accornero has argued, in an attempt to explain why Taylorism developed in the US rather than the UK, that the prerequisite

[13] On ENIOS see G. Sapelli, *Organizzazione, lavoro e innovazione industriale nell'Italia tra le due guerre* (Turin, 1978), and also P. Fiorentini, 'Ristrutturazione capitalistica e sfruttamento operaio in Italia negli anni '20', in *Rivista storica del socialismo*, 30 (1967).

[14] See D. Bigazzi, 'Organizzazione del lavoro e razionalizzazione nella crisi del fascismo 1942–3', in *Studi storici*, 2 (1978). Bigazzi does, however, argue that the work of ENIOS in this period did have a lasting influence after the fall of Fascism and the Liberation, when this organization reappeared under another name, this time with a more openly American character.

[15] Ibid. 370.

conditions for the introduction of scientific management are weak unions and a good supply of unskilled workers.[16] According to this argument, Fascist Italy with its puppet unions and the mass migration of rural workers towards the urban areas in defiance of the Fascist migration policy appeared, to a certain extent, an ideal situation for the introduction of scientific management. But the Italian example seems to undermine Accornero's thesis, since, despite the wealth of writings which appeared on this subject,[17] the apparent enthusiasm for Taylor and Ford was only poorly matched by concrete experiment in Italian industry. The reasons for this, as Sapelli points out, must be sought in the particularly backward structure of Italian industry in this period:

> Indeed the situation in Italy was far from providing fertile ground for the spread of scientific management methods. There were numerous problems. One was the fact that there had been relatively little industrial concentration. Furthermore, the expansion of the type of industries suitable for the introduction of mass production methods had been limited (and was destined to remain so due to the small size of the internal market and the crisis in international relations). Other problems included the growth in parasitical incomes, increasing speculation, and the existence of large numbers of small and medium-sized firms still using backward production methods and unable to modernize due to their extremely diversified product ranges. These represent only a few of the problems which had to be solved before a large-scale introduction of scientific management methods was possible. The revaluation of the lira which had compelled firms in every sector to cut costs, and the problems of the financial market which weighed heavily on smaller firms, only served to worsen this situation.[18]

Only in some of the larger Italian firms was it economically viable to introduce American management methods. In the majority of Italian firms, especially the small and medium-sized ones, Taylorism and Fordism had little or no impact. Even in many large concerns production methods barely changed. Most industrialists, for example, displayed very little interest in the international standardization campaign,[19] and in many firms

[16] Accornero, 'Dove cercare le origini del taylorismo'.
[17] Sapelli, *Organizzazione*, passim.
[18] Ibid. 122.
[19] On the early emergence of standardization see Noble, *America by Design*, ch. 5.

only a severely modified form of Taylorism was introduced, which concentrated on the cutting of costs through the raising of output per worker.[20] Rationalization in the majority of Italian firms tended to mean simply the use of bonus incentive systems such as the notorious Bedaux and the speeding up of the work tempo, rather than a more far-reaching reorganization of production methods. There was, admittedly, some technological innovation, mostly in the form of automatic or semi-automatic machine tools, thereby enabling a reduction of the workforce and a type of mass production, but this new technology was not accompanied in most firms by actual organizational changes.

In the textile industry, for example, which was still one of Italy's largest manufacturing sectors in this period, the restricted size of the internal market made the wholesale introduction of scientific management methods impossible, but the revaluation crisis of *Quota Novanta* in the later 1920s meant that costs had to be cut drastically. This was achieved largely by the introduction of automatic machinery and a dramatic speeding up of the pace of work using bonus incentive pay systems. Productivity was often raised simply by making each worker operate more than one machine, and rational methods of work organization, which were being tried in textile industries in some other countries, were hardly introduced at all. The result of this situation was that conditions in the Italian textile industry were very bad, giving rise to strikes and unrest in spite of the political repression.[21]

Only in a handful of large industries was there a serious attempt to implement Fordism and Taylorism.[22] Of these, probably the two most 'Americanized' were Fiat[23] and Magneti Marelli, reflecting the similarity of their products, which were easily standardized and mass-produced. It must be stressed, nevertheless, that the introduction of scientific management

[20] Sapelli, *Organizzazione*.

[21] See B. Bianchi, 'I tessili: lavoro, salute, conflitti', in *Annali Feltrinelli* (1979–80).

[22] e.g. the Olivetti typewriter factory at Ivrea. See G. Pedrocco, *Fascismo e nuove technologie: l'organizzazione industriale da Giolitti a Mussolini* (Bologna, 1980), 67–70; A. Olivetti, 'L'organizzazione in una fabbrica italiana di macchine per scrivere', in *L'organizzazione scientifica del lavoro*, 10 (1928).

[23] D. Bigazzi, 'Gli operai della catena di montaggio: la Fiat 1922–1943', in *Annali Feltrinelli* (1979–80).

methods to some extent represented a deliberate strategy and was not simply made inevitable by the nature of the product,[24] as is evidenced by the fact that in other car firms, such as Alfa Romeo, far more traditional work organization methods were maintained.[25]

In inter-war Milan, Magneti Marelli seems to have been the only large firm to undertake a major reorganization of production making use of the ideas of both Taylor and Ford. Furthermore, in line with many American companies who by this period had become very critical of Taylor's original methods, Magneti displayed a great deal of interest in the ideas of the so-called 'revisionist Taylorites' who stressed the importance of 'human engineering' (see Chapter 3). As such it became, in many ways, a 'model factory', the focus of innumerable visits by Fascist hierarchs, foreign dignitaries, and, on more than one occasion, Mussolini himself.

'SCIENTIFIC MANAGEMENT' AT MAGNETI

Production at Magneti was labour intensive and this, probably more than any other factor, accounted for the firm's interest in American ideas on scientific management. There was also the stiffness of foreign competition for a firm that had developed as a war industry. Production had to be economic on world markets if the nascent Italian magneto industry was to survive, since in the 1920s the market was dominated by efficient foreign firms such as Bosch, itself in the forefront of the introduction of scientific management methods in Germany.[26] Furthermore,

[24] David Noble makes the important point that the introduction of 'modern management' methods as a whole needs to be recognized as the deliberate work of people (engineers in particular), not simply the logical outcome of technological change; technological development itself, as he rightly points out, being in itself a human creation (*America by Design*, 258).

[25] D. Bigazzi, 'Management strategies in the Italian automobile industry', in J. Zeitlin and S. Tolliday (eds.), *Between Fordism and Flexibility: The International Automobile Industry and its Workers* (London, 1986). See also Bigazzi, *Il Portello. Operai, tecnici e imprenditori all'Alfa Romeo 1906–1926* (Milan, 1988).

[26] H. Homberg, 'Anfänge des Taylorsystems in Deutschland vor dem Ersten Weltkrieg: Eine Problemskizze unter besonderer Berücksichtigung der Arbeitskämpfe bei Bosch 1913', in *Geschichte und Gesellschaft*, 2 (1978).

products manufactured by this sector were generally small and suitable for standardization; many of them, such as car parts, were destined to be sold to firms like Fiat, themselves involved in mass-producing standardized products requiring standardized parts. Indeed the firm's close links with Fiat were very probably a crucial factor in the push to modernize production methods. Also important to the introduction of these ideas at Magneti was undoubtedly the rather cosmopolitan background of the Quintavalle brothers, who had travelled and even worked abroad.

It seems certain that the Magneti managers had actually read the works of Frederick Taylor and his followers (the so-called Taylorite engineers) and probably also those of Henry Ford,[27] but their knowledge of American management systems was not based on reading alone, as the Quintavalle brothers themselves, especially Umberto the engineer, visited American firms on a number of occasions and, from the early 1920s, managerial staff were sent to America on fact-finding trips to visit 'rationalized industries':

> some of our managers were especially sent to the 'school of industry' as America has been called. There they visited all the firms in our sector and then returned to Italy, bringing with them . . . a clear idea of the potential for the introduction of scientific management in light engineering. The complete reorganization of our production methods dates, in fact, from this period.[28]

Whatever the origin of their ideas, the Quintavalle brothers set out, from the founding years of the new firm, to 'rationally organize' production, combining more 'classical' Fordist and Taylorist methods such as new technology and a rationalized layout with an emphasis on the importance of the 'human factor' in production following the school of revisionist Taylorites, who had already begun to criticize Taylor's original ideas severely by this period.

[27] Taylor's work, *The Principles of Scientific Management*, was translated into Italian first in 1915. The first Italian translation of Henry Ford's book *My Life and My Work* was published in 1925. It seems highly probable, in any case, that the Quintavalle brothers read in English due to their extensive travel abroad.

[28] 'Dalle origini ad oggi', *SB* 9 (1929), 10.

MACHINE TOOLS

One aspect of this rationalization was the fact that, as in many other firms in the 1930s, Magneti installed individual engines for its machine tools, replacing the old transmission belts. This made the shops cleaner, quieter, and safer and enabled machines to be transferred more easily from one shop to another. It also saved the endless delays occasioned by a breakdown of a transmission belt or the central engine. Far more important, however, was the introduction of a large number of new automatic machine tools such as turret lathes and automatic drill presses, which were introduced into many shops replacing manually operated machines.

The machines being introduced were mostly multiple function machines which could carry out a sequence of operations according to preset patterns. Some had one particular pattern already built into the mechanism, whereas others were more flexible and could be adapted to a series of different operations. Typical of these machines was the automatic turret lathe which was introduced widely at Magneti. This had a series of tools carried in a revolving turret. After each tool completed its job the turret turned automatically to bring into play the next tool required in the sequence.

The importance of these machines was that they could complete a series of operations quickly, precisely, and automatically, thereby making it easier to manufacture completely standardized products, cutting work times drastically and in places enabling skilled craft workers to be replaced with unskilled or semi-skilled workers who were much cheaper to employ. Many of these machines could work on a number of parts at once, although only one operator was necessary.

For example, in the late 1920s the firm began to use the continuous loading Loewe vertical spindle milling machine fitted with special tooling. This could carry out the complete milling of magneto bases working on eight parts at once, and the time needed to complete the operation on each one was cut from two-and-a-half minutes to forty-five seconds. The work of the machine operator was rendered relatively simple, being mainly the loading and unloading of the machine. The factory newspaper presented this dramatic deskilling of previously skilled work as a benefit to the work-force, as less physical effort was required:

productivity rose from twenty-four to eighty parts per hour. In other words, output was tripled without the need for an increased number of machine tools or greater exertion from each worker. On the contrary, the worker now actually has to put in far less effort.[29]

This example is from the 1920s, but the introduction of automatic machinery seems to have been continuous throughout the inter-war period as production grew, making investment in new machinery possible. Gradually more and more previously manual jobs were automated, as this description of the Spark Plug Shop, published in the factory newspaper in 1936, emphasizes:

A battery of automatic machine tools provides a continuous supply of semi-finished parts, which, after passing through an initial quality control check, flow on to the finishing lines. Due to our large output, we have been able to equip these machines to carry out a single task each. We use the machinery for a number of different jobs such as thread cutting, soldering the central nickel electrodes to the shaft, preparing the electrodes, and finishing the various components. These machines send a rapid stream of finished parts to the final quality control check . . . This high level of automation ensures that production is both rapid and precise.[30]

The great extent to which automatic machines were introduced in this firm is confirmed by a report on the factory made by the government's secret industrial inspectors, the *osservatori industriali*, which listed the machines used in the plant in April 1937. There were, for example, about 140 turret lathes and a large number of other types of automatic machinery. In the Screw Cutting Shop alone, out of 151 machines, 57 were various types of turret lathe (Foster, Pittler, and Becker), 30 were simply described as 'automatic machines', and quite a number of the other machines were at least semi-automatic.[31]

[29] *SB* 4 (1927), 21. Another example was a new type of drill used in the production of magnetos, which replaced 4 old drilling machines and cut processing times per unit from 2 minutes 25 seconds to only one minute (*SB* 8 (1927), 21).

[30] *SB* 5 (1936), 17.

[31] ACS, OI, b. 22, f. 22 'FIMM', report by 1° Cap. Spadaccino Arturo (Osservatore compilatore 203E) Apr. 1937. 'Impianti e macchinari esistenti nello Stabilimento della Fabbrica Italiana Magneti Marelli.'

Many of these machines were initially imported from abroad (mainly from the US and Germany) and members of the management went especially to choose and buy them. These imported machines were usually adapted in the Machine Tool Room. Other machines were designed and built by the firm's engineers, sometimes following the suggestions of foremen or workers.[32] From 1940 Magneti set up its own subsidiary company in nearby Caravaggio, the Sezione Autonoma Macchine ed Attrezzature Speciali (SAMAS), which both devised and adapted machine tools and also studied imported machine tools in order to modify or copy them.

QUALITY CONTROL

An integral part of Magneti's approach to the organization of production was the introduction of a series of quality control checks. These were necessary since, without them, the speed-up of the work-pace would simply lead to an increase in the numbers of faulty unsaleable products:

> at most stages of the production cycle each operation is immediately followed by a quality control check. Like production methods, inspection itself has been mechanized and these controls, apart from making work smoother in the shops themselves, ensure that the assembled products arriving at the final quality control check have a good chance of being perfect as they are composed of perfect components. By identifying faults as soon as they actually occur we can eliminate problems immediately and avoid the time-wasting work of dismantling and repairing defective goods.[33]

In more 'traditional' firms quality control had been the responsibility of the foremen, but in the Taylorized firm with 'functional foremanship' this was done by specialized personnel.

Many of the quality control machines were actually built by the firm itself. The inspection network was costly to install, and it was not created immediately but gradually evolved over the

[32] Many American firms used this system of 'suggestion boxes', a tacit acknowledgement that management could not really expropriate all of the workforce's skill and knowledge. (See Noble, *America by Design*, 267).

[33] *SB* 5 (1936), 17.

inter-war period. In the early 1920s, for example, some of the intermediate inspection work was still being done by the heads of shop,[34] but as the economic recession began to bite, this became seen as increasingly important to cost cutting:

As far as cost cutting is concerned . . . labour costs can be reduced by introducing intermediate quality control checks at crucial points on the assembly lines, in order to virtually abolish the final inspection, since it is only through repeated intermediate checks that we can hope to produce perfect goods.[35]

By the late 1930s the number of specialized personnel involved had become very large. Each section of the factory had a head quality control inspector, and under him (they were all men) were intermediate technical staff who moved around the shop supervising the workers who did the routine inspection at fixed points on the production lines. All the quality control operations were co-ordinated by the central Quality Control Department. To give some idea of the size of the quality control operation it is sufficient to note that, according to the factory newspaper, by 1936 there were 150 inspectors on the shop-floor and these were supervised by 25 intermediate inspectors. There were also quite a number of workers employed in the Special Testing Room and in checking materials as they came into the factory and doing the final quality control checks on the finished products.[36]

FACTORY LAYOUT

From its earliest days, the new A Plant was laid out according to so-called 'rational' criteria:

The factory, which is organized according to extremely modern criteria, is divided into a series of shops, each of which is devoted to producing one or more separate parts of the magneto, so that each process follows its normal and logical cycle. In other words, we have abandoned the out-of-date idea that each shop should contain one type of machinery only. Each of our shops contains all the machinery necessary for the

[34] 'Il pelo nell'uovo', *SB* 5 (1924), 11.
[35] *SB* 11 (1930), 85.
[36] 'La qualità nella produzione', *SB* 6 (1936), 30.

production of one particular part, arranged so that the raw materials or components arrive at one end, move continuously from machine to machine and from bench to bench according to the needs of production, so that at the end of the line a completely finished part is ready to leave the shop.

In this way all non-essential movements of materials are avoided and it is possible to introduce mass production, so that each worker can remain in one place and receive the part from the previous worker, carry out one particular operation, and no other, and then pass it on to the next worker further down the line.[37]

As this extract, written in 1924, demonstrates, from the beginning the new firm had a carefully planned 'modern' layout.[38] Throughout the 1920s and 1930s, this was modified to a certain extent to meet increasingly complex production needs, but the basic concept of organizing the shops according to the products they made (radios, magnetos, windscreen wipers, etc.), rather than by which type of machinery they required, remained. There were also a few shops which produced components required in the other parts of the factory (Shearing Shop, Foundry, Screw Cutting Shop, Winding Shop, etc.). These more specialized shops were largely located on the ground floor of the factory together with the 'service' departments such as the Tool Room. Most of the store rooms and the research laboratories were in the basement. The shops for processing and assembly were all situated on the higher floors of the factory, so that generally everything in this plant flowed upwards. Materials came into the general stores in the basement, were then processed on the floor above, and the resulting components were moved up to the higher floors of the factory to be assembled into finished products.[39]

[37] *Nei campi della gloria* (Milan, 1924).

[38] For maps showing the layout of the A and B plants see appendix 3 in P. R. Willson, 'Women Workers, Scientific Management and Workers' Welfare: The Magneti Marelli in the Fascist Period', Ph.D. thesis (University of Essex, 1986).

[39] This upward-flowing design was similar to the layout of the Lingotto Fiat plant, which even included a test track on the roof for the final product. The Lingotto plant, which was modelled on the lines of Ford's Highland Park, had five floors and the engine and chassis 'grew' from floor to floor in two separate areas of the building until they flowed together for the final assembly on the fifth floor. The vehicles were then taken up to the test track which was situated on the roof of the building. (See Bigazzi, 'Gli operai'.) Because of the relatively

To move parts and materials around the factory and create this upward flow, increasingly sophisticated internal transport systems were introduced. Aerial transportation was used to carry materials and components from one shop to another throughout the factory, and in some shops gravitational systems moved things around slowly at a fixed pace. Increasingly materials were distributed from a central stockpile and intermediate stocks were eliminated. In 1925 Magneti became the only Italian firm apart from Fiat to introduce conveyor belts for assembly work on the Fordist model.[40] It is difficult to exaggerate the difference that conveyor belt production meant to the worker. Management control of the pace of work was now no longer simply a question of timing individual tasks, but work became subjected to the general relentless rhythm of the moving belt:

The materials are brought in on special trolleys and distributed according to a specific order along a counter from which they are fed onto the assembly line. This counter is the first interesting feature of the Assembly Shop. After careful study it was decided to construct this so that the top part consists of a set of shelves and the lower part of drawers. This system makes it possible to continuously supply the line whilst simultaneously carrying out a preliminary inspection of the materials.

The counter is positioned at the head of the conveyor belts so that the materials can be placed on this according to the order of assembly.

Thus the workers receive a continuous supply of the parts they require and with the same methodical, identical movement, they repeatedly carry out the same task again and again.[41]

The factory newspaper positively enthused about the productivity potential of this new equipment:

the time required by the worker to finish each task before the conveyor belt carries the part on is carefully calculated and the use of pneumatic cleaning machinery etc. all ensure that the worker does not have to

smaller size, greater variety, and simplicity of the Magneti products compared with the whole vehicles produced at Fiat the layout needs were somewhat simpler.

[40] Some other firms apart from Magneti and Fiat, such as jam manufacturers and biscuit packers, subsequently introduced conveyor belts in the interwar period, but these remained rare. (Sapelli, *Organizzazione*, 244).

[41] 'Nulla dies senza linea', *SB* 11 (1924), 12–13.

waste a second . . . Recent highly interesting studies have revealed the amount of time a worker wastes simply in picking up the part from a table next to him or from the hands of another worker, or in picking up the necessary tools etc. It is only a matter of seconds but added together they represent a huge waste of the productive potential of each worker and they tire him pointlessly.[42]

This initial experiment was built upon and Fordist-style conveyor belt assembly was introduced for practically every product, and moving belt transport systems began to be used even in other parts of the factory, such as in the storeroom for semi-finished parts.

Thus, by the 1930s, the physical movement of materials, semi-finished parts, and products was almost all mechanical, with aerial transport systems carrying them from shop to shop and conveyor belts linking machines where each specialized process was carried out repeatedly. Apart from speeding up production times and cutting down on labour this system meant that stockpiling was kept at a minimum and capital investment reduced.[43]

In the B plant the work cycle was also organized so that parts 'flowed' from shop to shop; but as this plant was far smaller and produced only a limited range of products, the layout does not seem to have been so distinctly organized along the 'flow' pattern as the A, probably because parts and raw materials could be moved around the smaller area far more easily. As in the A Plant, some of the shops such as the foundry were organized according to the type of machinery they contained. There were also a number of assembly shops each producing a specific product such as submarine batteries.

From the early 1940s shops began to be transferred from the A

[42] 'Nulla dies senza linea', *SB* 11 (1924), 12–13.

[43] '. . . we have managed to reduce the amounts of raw materials in circulation relative to the number of parts produced by about one quarter, thereby enabling a saving of about a million lire in capital investment. This has been achieved by a careful analysis of the work process to make it continuous and by eliminating all intermediate stockpiles in the storerooms or next to the machinery through the use of mechanized internal transportation systems. In the Assembly Shop, for example, where we have introduced a Fordist-style conveyor belt, we have both reduced the workforce and also cut the number of components in circulation from those needed for 3,000 final products to only 450.' (U. Quintavalle, 'L'organizzazione razionale della produzione', *Realtà* (1 Mar. 1927), 238).

in Sesto to the larger, more modern N plant in nearby Crescenzago. Basically, although the new plant has been depicted as very different from the older A,[44] its layout might be more accurately described as an extension of the reorganizational trends in the A plant. The difference was that the new plant was designed from the start with a so-called 'rational layout', unlike the A where the transportation of parts was rendered complicated by the design of the building which had been added to throughout the inter-war period in a piecemeal fashion. The new plant had only one storey, and 'the assembly lines were expanded and integrated in this plant, which was laid out horizontally, making it possible to expand or contract each section at will'.[45] In this new plant some of the 'inconsistencies' of the A plant were smoothed out and, for example, much of the work of the Winding Shop was divided up amongst the various production lines. In the N the general shops which played a part in the manufacture of all types of products were situated at the beginning and the end of the lines for each individual product. Raw materials were processed first by a series of shops such as the Foundry and the Shearing Shop. After the assembly lines came the testing and packaging departments.

A 'CLOCKWORK FACTORY'?

The 'flow' layout of the factory floor, coupled with the mechanized movement of materials, gave Magneti almost the aspect of a 'clockwork factory' where the machine ruled, as this extract from the factory newspaper implies:

> There are no useless movements or bottlenecks in the supply of materials. The transporter shunts each finished part to its destination and with its regular movement imposes the work tempo for the entire shop.[46]

Perhaps the best way to look at how the 'clockwork factory' functioned is to follow the manufacture of one product: radios.

[44] *Milano tra guerra e dopoguerra* (Milan, 1979), 107.
[45] F. Guizzi, 'La Fabbrica Italiana Magneti Marelli', in *La ricostruzione nella grande industria*, 289.
[46] 'Il fattore logistico nella produzione', *SB* 4 (1936), 17.

First came the work of shops such as the Shearing Shop and Screw Cutting Shop which made all the small components necessary to the whole product. These were then moved upstairs and assembled into parts such as the base and the loudspeaker. After the sub-assembly, the semi-finished parts were channelled into a special store where all the components from screws to loudspeakers essential for the final assembly were kept:

> The function of this stockroom, positioned at the head of the assembly lines, and linked to these by a conveyor belt, could be roughly compared to the reservoir of a plumbing system. Indeed, the stockpile of parts kept in the stockroom at all times ensures that the assembly lines never have to be stopped because of problems which inevitably arise in the other shops of the factory, and enables production to be continuous.[47]

After this, the parts were assembled on two successive assembly lines, one for the chassis and the other for the finished product. All the movement of parts from shop to shop was done by means of overhead aerial transport systems or conveyor belts.

The factory was not, of course, really automatic, but the product of a huge amount of planning. An extensive structure of organizational departments enabled the management and most particularly the technical management to take over control of the work of the factory, 'relieving' the work-force of much of the thinking work: the essence, in fact, of Taylorism.

PLANNING AND CONTROL

The mental work of production was centralized under the control of this technical management, which became the 'brains' doing the thinking for the 'hands'—the workers and the machinery. The factory newspaper was not shy to point this out:

> As in all the best foreign firms, the mass production shops have the sole, exclusive task of producing in a certain way a certain predetermined quantity of products, without having to worry about either the way

[47] 'Come si construisce un apparecchio radio', *SB* 2 (1937).

work is organized or how to obtain the necessary tools etc. All this is taken care of by special departments created for the purpose.[48]

In the early 1920s production was already being organized, even though the firm was still small, by a series of special departments. The Works Technical Department drew up special worksheets (*diagrammi di lavorazione*) for each process, noting the type and sizes of machinery and tools to be used, the work times, and so on. The Production Department drew up monthly programmes and used these to ensure that each shop was supplied with the correct materials and tools needed to carry them out. There was also a special department to take care of the standardized measuring tools and a sample of each product which was being made. A further department distributed tools and machinery to each shop. By the early 1920s, therefore, the factory had already organized itself to make the thinking work largely a management prerogative and was at least in theory capable of manufacturing standardized mass products according to a strictly defined management plan. The management claimed that this was the key to dramatically increased productivity.[49]

Throughout the 1920s and 1930s innovations continued. As the management became more experienced and the logistics of production increased enormously with the growth of the firm and the much wider range of products, the firm developed an increasingly sophisticated organizational hierarchy to orchestrate the functioning of production. By 1928 the system of developing a *mappa di lavorazione* had been introduced. This described the materials, machinery, tools, and labour needed for each product. From this the production cycle was worked out and the *mappa* described in detail each specific process which the raw materials had to undergo and the movement of materials from shop to shop and from machine to machine. The *mappa* was also used by various other departments to calculate times, the standard tools

48 *Nei campi della gloria*, 6.

49 'Only two years after we first began to introduce rational organization methods, with the introduction of the analysis of the work cycle, a linear layout, standard classification for equipment and tools, and the centralized control of production, without buying any new plant, we increased our production by a factor of almost four, even though the workforce was only expanded by about ten per cent' (Quintavalle, 'L'organizzazione razionale', 237).

necessary, and the materials which needed to be bought. The actual carrying out of the work described by the *mappa* was supervised by the Production Management, who worked according to monthly plans.

By 1936 this whole process had been further refined. By this date an Organizational Department had been created. Once the General Management had decided to go ahead with the manufacture of a certain product the Organizational Department drew up a *sviluppo della lavorazione*, a more sophisticated version of the previous *mappa*. This

> analyses the product part by part, determines exactly how many of each component and which raw materials are needed, defines the succession of operations to be carried out as the assembly progresses, lists the machinery and tools necessary, and finally calculates the time required for each of these operations.[50]

Once the *sviluppo* had been issued various departments automatically calculated the requirements of labour, tools, machinery, and materials and ensured that these were ready for production to commence. The Sales and Marketing Department drew up a monthly programme for output and issued the order for production to begin. It also issued each shop with schedules detailing the monthly output for each product.[51]

The day-to-day running of production on the shop-floor was the job of the staff responsible to the planning departments of the factory. They kept a detailed record of output so that this could be watched and co-ordinated from the central departments:

> the Heads of Shops must constantly supervize workers to advise them and help them to overcome technical problems as these occur. In this way the Production Management Office, through the special organization of 'product foremen' (one for each product), can guide and supervize how production is progressing, ensuring that raw materials are not lacking, and that the production of components and their

[50] *SB* 4 (1936), 20.

[51] A Magneti engineer, Vico d'Incerti, compared these to the music of an orchestra: 'The "*tabelle d'officina*", to use an analogy for those who don't understand these things, are similar to the various "parts" of an orchestral score. Preparing them is like a composer deciding which instrument is to play each part, when each instrument comes in, the timing, etc. These are all meticulously calculated and it is a serious matter if each note is not played at exactly the right moment. If anyone plays out of time the overall effect is ruined.' (Ibid. 21).

movement from shop to shop occurs strictly according to the predetermined programme. They also have to sort out problems and be prepared to make timely changes of plan which may become necessary if insuperable problems crop us such as raw materials not being delivered on time, crucial technical variations becoming necessary, unforeseen build-ups of stocks, etc.[52]

To ensure that the sub-assembly was adequately supplied it was enough to ensure a good supply of raw materials, but, in order to keep the final assembly functioning smoothly, an intermediate stockpile was needed. A great deal of attention was paid to the question of stockpiling and a complex system of records was developed to ensure that excessive stockpiling of parts or materials did not occur, since this necessitated unnecessary investment. Provision was also made for the reintegration of faulty parts which had been sent back by quality control, and the plan could also accommodate technical variations in production.

By the 1930s, therefore, a complex hierarchy of differing planning departments had been developed to coordinate output. The thinking work of production had been largely divorced from its execution, and control was centralized, thereby deskilling much of the work of the foremen and greatly increasing the power and responsibilities of the engineers. The firm was organized in this way from the early 1920s, but, as production expanded and became more complex, the system was refined and enlarged.

RESEARCH AND DEVELOPMENT

In Giolittian Italy there had been few links between science and industry, but with the advent of the First World War more attention was paid to research and development, with a few firms such as Pirelli setting up research laboratories. Many of the scientists who had become involved with industry in this period maintained the link after the end of the war, especially in the chemical and electrical industries. The Fascists declared themselves in favour of fostering greater co-operation between

[52] Ibid.

science and industry, although not very much was actually done to this end; much of the Fascist push for the encouragement of applied science was little more than rhetorical propaganda. Nevertheless, in this period greater links were created between science and industry and, especially in the 'autarkic' period, some of this was due to the efforts of the regime. But by the end of the 1930s Italy still lagged far behind the other industrialized nations[53] in this field, and generally only a few large firms really had important research and development programmes.[54]

In this context it is clear that Magneti had one of the most energetic approaches to research and development in Italian industry. Already in the 1920s a degree of attention was paid to this question, which was of vital importance to a firm operating in a competitive international market, and a number of research laboratories were installed. By the late 1920s, for example, there was a Study Department which 'designed' each product according to the specifications given it by the Sales and Marketing Department. The design was then made up into samples by the Testing Shop. A Scientific Laboratory studied their functioning and suggested modifications. In the Special Testing Room (which the firm claimed to be the first of its kind in Italy), products could be practically tested for resistance to factors such as cold, vibration, and altitude. This room was equipped with special machinery to simulate low pressure, extreme cold, and so on.

In the 1930s the firm developed a number of scientific research laboratories to design, test, improve, and update products and materials. This helped it bring out a constant flow of new products. By the late 1930s the original, general scientific laboratory of the 1920s had been split into a series of specialized ones each with a number of scientific personnel. In the A Plant by 1938 there were twelve research laboratories, whose work ranged from the testing of raw materials to the designing of the

[53] For a critical evaluation of the growing links in the USA between academic scientific research institutions such as MIT and industry see Noble, *America by Design*, ch. 7. According to Noble, by the eve of the First World War science was being 'pressed into the service of capital' (p. 147).

[54] R. Maiocchi, 'Scienza, industria e fascismo (1923–1939)', *Società e storia*, 2 (1978).

first television sets.[55] Also involved in research and development in this period were departments to design new products, a Technical Department, a Standardization Department, a patents office, and a shop for preparing samples of each new product.

Apart from the firm's own technical staff who worked in these laboratories there was also a certain amount of co-operation with academic scientists in universities. In 1938, shortly before he received the Nobel Prize for Physics, Enrico Fermi was appointed to be head of the entire research programme. But Fermi's contribution to Magneti was to be short-lived, for soon after he had to flee Italy to escape the new antisemitic legislation.

The appointment of such a prestigious scientist indicates the firm's pressing need for good research and development in the period of the Fascist autarky campaign. According to the factory newspaper, Magneti threw itself into this campaign with the greatest vigour, even though (as was described above) Quintavalle himself did eventually come to feel critical about this particular aspect of government policy. But this did not prevent the firm from reaping the maximum publicity from its efforts, and the Magneti subsidiary FIVRE, for example, was constantly portrayed as being in the forefront of the drive to autarky, since radio valves had previously all been imported. Furthermore, as the Second World War loomed, self-sufficiency in raw materials seemed increasingly important and research for autarkic materials and machine tools became a focus for the firm's energies. The management was keen to project an image of itself as patriotic and autarkic and this became a central theme of the firm's advertising campaign, and numerous articles in the factory newspaper in this period glorify the firm's efforts to discover substitute materials which could replace imports.

[55] 'I nostri laboratori e l'autarchia intellettuale della Nazione', *SB* 5 (1938). The 12 research labs were: raw materials testing, scientific ignition parts designs, high-altitude testing, radio, short-wave, television, acoustics, radio receiver designs, radio transmitter designs, acoustic designs, chemical lab for use in radios, measuring and testing equipment. There were also a 'special testing room' and a 'special testing room for radios'. In 1939, furthermore, a special 'lab for designing radio equipment for the armed forces' was created as the firm prepared for war.

CLERICAL WORK

It is not easy to obtain information on the organization of clerical work. The factory newspaper, always keen to herald new developments on the shop-floor, is more reticent about the work of the offices. This was, nevertheless, undoubtedly a period of change in office work. This was largely in the form of technological innovation with the increasing use of office machinery such as typewriters and adding machines. As early as 1929 various types of imported adding machines were in use, and clerks had to be recruited from particular schools which gave training in their use.[56] Probably the most significant change was the introduction of American IBM calculating machines in about 1939 or 1940, to mechanically produce standardized pay slips. They radically changed the work done by many clerks, reducing some of them to doing simple card punching, a repetitive, monotonous task which was allotted entirely to women.[57]

Perhaps the most noteworthy aspect of clerical work in this firm is the sheer number of clerks, swollen by the complex administration involved in 'modern management'. By the end of 1942 the clerical staff were a staggering fifth of the entire work-force.[58] The percentage of clerical workers seems to have grown, as according to an article published in the Communist newspaper *Unità*, in 1926 only one in seven were white-collar.[59]

THE LIMITS TO TAYLORISM

This chapter has concentrated so far on looking at the ways in which Magneti was more 'advanced' in the introduction of Ford's and Taylor's ideas than the vast majority of Italian firms in this period. Without trying to deny that this was so, it is nevertheless true that Magneti did not actually introduce these

[56] Testimony of Maria B. (former clerical worker).
[57] Ibid.
[58] ACS, OI. b. 22, sf. 'FIMM: personale in forza al. 31. 12. 42'.
[59] 'Dalle Magneti Marelli di Sesto San Giovanni e di Canonica d'adda', *Unità* (11 Apr. 1926), 3.

ideas wholesale. This was, of course, true even in the home of scientific management, the United States, where very few companies really introduced a carbon copy of Taylor's ideas.[60] One reason may have been the amount of time necessary really to learn how Taylorism worked. Taylor himself calculated that two to three years were probably necessary for managers to acquaint themselves properly with his ideas.[61] Added to these problems were others for Magneti, such as the limited size of the Italian market and the problems of recruiting suitable technical personnel.

Most Magneti management publications of the period such as the factory magazine are characterized (particularly in the 1920s) by an endless gushing euphoria about American methods. But occasionally even the management was forced to admit candidly that not all was possible. As Michele Soliveri, the works manager of the A Plant, noted in 1927:

> It is clear that American organizational methods cannot be simply blindly imported as they stand into Italy. There, before an industry is set up, everything is prepared beforehand, and a product is chosen which is likely to have a large market. In Italy it is a different matter, as it is impossible to count on a really large amount of sales for any product, given the limited demand.[62]
>
> The truly scientific criteria followed by industrialists in the United States would be simply an unnecessary luxury for us.[63]

Added to this was the fact that in practice the difficulties of implementing these ideas were far from slight, and oral testimonies suggest that the introduction of many of the much

[60] An article published in 1915 reported that in this year only 1% of all American factories with over 100 wage earners had actually introduced scientific management. See C. B. Thompson, 'Scientific Management in Practice', *Quarterly Journal of Economics* (1915).

[61] As E. Layton has commented: 'Taylor was undoubtedly correct in his insistence that the technology must be learned at first hand and be embodied in a person in order to be diffused. And his estimate of 2 to 3 years for such a task was probably correct. But Taylor was never able to find a foreign engineer willing to spend that much time learning his system.' (Layton, 'The Diffusion of Scientific Management and Mass Production from the U.S. in the Twentieth Century', *Proceedings of the 16th International Congress in the History of Science* (Tokyo, 1974), 379).

[62] M. Soliveri, 'Venti anni di lavoro. Un periodo—un ricordo, 5 nov. 1907–5 nov. 1927', AISRMO, CLN, Carte Mantovani, p. 30.

[63] Ibid. 34.

vaunted management innovations went far from smoothly. For example, skilled worker Gaetano commented:

We always had lots of problems as regards the quality of production. When Magneti Marelli started to make its first products using American licences, well, there were always enormous problems. Because although we received the parts, the components . . . initially . . . from America, we had difficulty . . . we didn't know how to put them together. Didn't know how to assemble them . . . So our production system was one that created a great deal of problems. Especially in the field of radios. Our first radio, the famous first radio, the Coribante . . . it was a disaster! There was so much ineptitude, both practical and scientific. Because in order to do what they did in America or in England we needed a kind of technical training that hadn't yet reached Italy.

The firm's commitment to Americanism also appeared to change with time. In the 1920s exports had to a certain extent tempered the problem of the small size of the Italian market, but the increasingly restricted possibilities for trade in the 1930s limited the scale of mass production and the rate of investment in new machinery and research. By the end of the 1930s the factory newspaper dedicated far less space to cost-cutting American methods and these appeared to become secondary to the drive to produce an autarkic product. Patriotism and autarky became the key words in a firm whose fate was so closely dependent on state orders and, increasingly, rearmament. This situation was voiced by Umberto Quintavalle in 1938 when he wrote of the firm's research and development programme that, although this was perhaps limited in comparison with similar firms abroad, he considered it well suited to Magneti's needs:

It is not without a certain amount of envy that we regard the huge amounts of money earmarked for research in large foreign firms and the impressive scientific progress which has resulted from this.

This is only possible, however, in a different situation and in a different market. It is important to know when to compromise and not to delude ourselves, as some have previously done, into thinking that it is possible to simply transport American management ideas wholesale into Italy.[64]

In fact, probably one explanation for this new attitude was the

[64] U. Quintavalle, 'I nostri laboratori nel loro concetto informatore', in *SB* 5 (1938), 7.

increasing protectionism of the 1930s, which meant that Magneti was no longer really competing with these foreign firms but operating in a safe market. Cost effectiveness became less important to the firm's growth than state orders and rearmament. On the other hand it is also possible that the Magneti managers increasingly had become convinced that the problems arising from too much American-style innovation outweighed their value and that this might not necessarily be cost-efficient. Thus it may be that in the late 1930s a progressively more prudent and selective approach was being taken.

Yet, was the heyday of Americanism really over at Magneti? Maybe this article of Quintavalle's was to a certain extent simply rhetorical toadying, inspired by the fact that the regime's zest for rationalization was waning distinctly in this period. The emphasis was on things Italian and not on foreign, imported methods and it could, therefore, have been in the firm's interest to minimize its debt to Taylor and stress its efforts for autarky. It is quite possible, as in most of the other Italian firms which had taken an interest in American methods,[65] that during this period Magneti continued to press on with the reorganization of production and the introduction of increasingly sophisticated scientific management methods. In fact, in the 1930s, in a great number of ways the firm built upon the reorganization which had taken place in the previous decade and the Taylorist and Fordist models remained the management's inspiration.

BEYOND TAYLOR

In spite of these limitations which meant that Magneti should not be seen as simply introducing American ideas wholesale, compared with the vast majority of Italian firms in this period Magneti Marelli was in the vanguard of the introduction of Taylorism and Fordism in Italy. The factory was laid out according to 'rational' criteria, there was a complex system of quality-control checks, and the thinking work of production was largely divorced from its execution. There was also considerable

[65] '. . . the crisis which ENIOS underwent in the 1930s was not paralleled by an analogous crisis in the actual introduction of scientific management methods in industry' (Bigazzi, 'Organizzazione del lavoro', 370).

technological innovation, including the use of conveyor belts and the introduction of numerous automatic machine tools.

This approach to factory organization, radically new in the context of inter-war Italy where production methods in many firms were far less sophisticated, was facilitated by a number of factors ranging from the type of product being manufactured to the attitude of the management. It resulted in a factory that seemed to function almost as if by clockwork, where the work many employees had to do was specialized and simplified. Production was dominated by the increasingly powerful engineers, who took over the thinking work, while many of the workers themselves were reduced virtually to the status of machines in the workplace, repeatedly carrying out the same narrowly defined small task all day long.

However, scientific management at Magneti was not limited to this rather mechanistic approach to the reorganization of production methods. By the 1920s modern management methods in the United States had already gone beyond this, and the work of the revisionist Taylorites had come to question many of the central tenets of Taylor's ideas, which were now seen as crude and unworkable. By the inter-war period the focus was on motivating the worker through incentives and so-called modern personnel management. In both these fields Magneti made a determined effort to be up to date.

3

'Taylorismo' at Magneti II: The 'Rationalization of the Human Factor'

THE introduction of Americanized production methods at Magneti was marked by considerable attention to 'the human factor', as the factory newspaper frequently termed it. In this respect the Magneti management was clearly in line with recent American innovations in management techniques. After the death of Taylor in 1915, a 'revisionist' form of scientific management had emerged in the US, with some of its leading advocates coming from the ranks of the early 'Taylorites'. The revisionists sought to remedy the most glaring problems produced by the implementation of 'pure' Taylorism. Certain of their innovations were simply an extension of Taylor's ideas, refining them and making them more 'scientific' (indeed, some of the Magneti innovations described in the previous chapter were inspired by the ideas of revisionist Taylorites), but others attempted to go further and tackle what was perhaps the most significant of the problems which had emerged, namely the extent of workers' opposition to Taylorism. As David Noble comments:

> The original scientific-management approach to human engineering . . . was less than adequate. Taylor and his associates essentially viewed the worker as a simple, and simple-minded, 'economic man'. Although they paid close attention to the details of his work, the Taylorites relied upon elaborate, though psychologically crude, incentive-pay schemes to motivate him, or upon the pressures of the labor market and the 'shop-disciplinarian' to coerce him into 'voluntary' cooperation.[1]

[1] D. Noble, *America by Design: Science, Technology and the Rise of Corporate Capitalism* (Oxford, 1977), 264–5.

Dissatisfaction with these early management techniques led to a focus on the importance of 'human engineering', which merged old-style paternalism with scientific management and the new science of sociology. The aim was to create a contented workforce: 'the worker . . . was to be encouraged, on the basis of "the most thorough comprehension of the human being" to play the game as management called it, to become a member of the team.'[2] Workers were to be made 'members of the team' in a number of ways. Some revisionists focused on the psychological underpinnings of workers' motivations whilst others sought to harmonize union–management relations through the new science of industrial relations. Many saw an important role for welfare facilities. These initiatives should not, however, be confused with old-style paternalism which sought primarily to defuse social tension through the alleviation of the suffering of the industrial worker, since the new approach now saw the worker primarily as an object of study and control. Furthermore, it was believed that this modern approach could boost profit by binding the worker to the firm and its productivity goals. It was argued, for example, that the inefficiency produced by a high rate of turnover (typical of many American firms) could be reduced. By the 1920s in America 'industrial relations' or 'human relations' was widely seen as important and many companies attempted to put some of these ideas into practice.[3]

These ideas spread to Europe, and some historians have argued that 'human engineering' had even more impact on management methods there than in the US due to the greater maturity of the European labour movement. This meant that employers had to pay greater attention to winning the consensus of their work-forces for the introduction of scientific manage-

[2] D. Noble, *America by Design: Science, Technology and the Rise of Corporate Capitalism* (Oxford, 1977), 268.

[3] Henry Ford, for example, expressed an enthusiasm for some of these new ideas, such as the importance of an industrial health and safety programme. He maintained that not only was it socially desirable to protect the health of the workforce but also that this could help keep productivity high. In his view, the social benefits of improved safety could be an integral part of scientific management: 'One point that is absolutely essential to high capacity as well as to humane production, is a clean, well-lighted, and well-ventilated factory' (H. Ford, *My Life and My Work*, new edn. (London, 1924), 113).

ment.[4] In Fascist Italy, however, since the unions had been destroyed, the spread of these ideas was more limited.

They were not, however, totally neglected. There was, for example, a certain amount of interest in industrial psychology, especially amongst academics such as Agostino Gemelli of the Catholic University of Milan[5] and Mario Ponzo of the University of Turin.[6] Numerous studies in this field were published in management journals, and, in November 1932, a free industrial psychology advisory service for managers was set up in Rome by ENIOS. But these attempts proved to be largely fruitless and only a handful of firms even attempted to apply such ideas.[7] With regard to the vast majority of Italian firms, Sapelli was probably correct when he asserted that: 'In spite of the remarkable efforts of men like Agostino Gemelli, there is no evidence of an effective introduction of industrial psychology ideas in industry.'[8]

Magneti Marelli's interest in revisionist Taylorism was certainly very unusual in interwar Italy and in a number of respects it can

[4] A. Pichierri, 'Diffusion and Crisis of Scientific Management in European Industry', in S. Giner and M. Scotford Archer (eds.), *Contemporary Europe: Social Structures and Cultural Patterns* (London, 1978), 60.

[5] See e.g. A. Gemelli, 'L'orientamento attuale della psicotecnica industriale e i suoi risultati', *L'organizzazione scientifica del lavoro* (1932), 63–4. Gemelli's research was carried out in a special psychology laboratory he set up in the Catholic University of Milan. See E. Cancellieri, 'Gli studi del laboratorio di psicologia dell'Università Cattolica del Sacro Cuore. Possibili sviluppi e riflessi sull'industria', *L'organizzazione scientifica del lavoro* (1932), 138. A number of publications have appeared on the work of this remarkable man, which covered a wide range of interests. See e.g. R. Zavalloni, *Padre Gemelli educatore sociale* (Milan, 1960); G. Rumi, 'Padre Gemelli e l'Università Cattolica', in *Modernismo, Fascismo, Comunismo* (Bologna, 1972).

[6] See e.g. M. Ponzo, 'Selezione e orientamento professionale nei loro rapporti e nelle espressioni di alcune odierne iniziative italiane', *L'organizzazione scientifica del lavoro* (1935), 11–14.

[7] G. Sapelli, *Organizzazione, lavoro e innovazione industriale nell'Italia tra le due guerre* (Turin, 1978), 361–2. Apart from Magneti the only example that I have found of a firm introducing an industrial psychology unit in the 1920s is the case of the mines in Gennemari Ingurtosu (Sardinia), where one was functioning by 1929. Here, as at Magneti, its role in fighting the accident rate was stressed. One reason why this firm actually introduced these ideas, rare in Italy, may be the fact that these mines were owned by a French company. See P. Audibert, 'Nota sul gabinetto psico-tecnico della Miniera di Gennimari Ingurtosu', *Resoconti dell'Associazione Minieria Sarda*, 8 (1929), 4–9.

[8] G. Sapelli, 'Formazione della forza lavoro e psicotecnica nell'Italia fra le due guerre mondiali', *Quaderni di sociologia*, 1 (1977), 21.

be considered a pioneer. Admittedly, some of its extensive welfare programme was essentially indistinguishable from classic paternalism, but other initiatives which are described in this chapter had a different inspiration. They were seen as an integral part of the management's frequent claim that scientific management could help worker and boss alike. As such their close relationship with scientific management practice was tirelessly stressed by management publications. However, as emerges from a closer examination of these policies, this was at a cost to the worker, for these initiatives extended management control and the hierarchical nature of the factory.

At Magneti, a belief in the importance of tackling the 'problem of the human factor' gave rise to a number of experiments and innovations. Of particular interest are the so-called 'scientific' recruitment procedures, the innovative organization of the Personnel Office, the workers' training school, and the health and safety programme.

TRAINING FOR TAYLORISM

The question of training was important in a rationalized industry since new types of methods required a 'new type of worker'.[9] A number of American 'scientifically organized' firms (including Ford Motors) had tried to remedy this problem by setting up workers' schools[10] and this example was followed by the two most 'rationalized' Italian firms, Fiat and Magneti.

Founded in 1925, Magneti's apprentices' school—the *Scuola Vedette*—offered technical and practical training to pupils of both sexes. The girls were all apprentice winders, whereas the boys

[9] For a general examination of the developments in technical education in this period and the attempts to adapt this to changing production methods see Sapelli, *Organizzazione*, 308–33, and F. Cereja, 'La scuola e il mondo del lavoro. Problemi dell'istruzione tecnica e professionale', *Annali Feltrinelli* (1979–80), 51–90.

[10] Even as early as 1902 General Electric had set up their own school at Lynn. Many other firms soon followed suit such as Westinghouse and Western Electric. In the US corporations not only set up their own schools but pushed for changes in the curriculum of state schools themselves to encourage more narrow vocational education which fitted schoolchildren to the needs of corporate enterprise. (Noble, *America by Design*, 302–4.)

were destined to work in a variety of roles within the factory as skilled workers, gang bosses, or foremen. A few even rose eventually to the technical staff. Pupils of both sexes also did sports and more general lessons. The facilities provided seem to have been good, and as such a useful showpiece for the newspaper, even keen to project a modern and efficient image. There was a properly equipped schoolroom and from 1927 a special Training Shop for the boys (girls being trained in the Winding Shop itself). During the part of the working day they were being trained, 'the pupils are paid . . . in exactly the same way as if they were working on the shop-floor'.[11] The *Vedette* also had to attend theory lessons after work, for which they were not paid.

In the early 1930s girls were excluded. It is unclear why this happened, as it does not reflect the deskilling of the winders' work since the mechanization of winding did not occur until the end of the decade. Girls continued to be trained in the shop itself as previously, but they were no longer called *Vedette*. They could continue with the domestic science classes which had been part of their training, but only as a 'leisure' activity.

In this period the number of male trainees increased gradually[12] and new specialized courses were set up for the retraining of skilled workers such as radio technicians. With the expansion of production for rearmament the training facilities were further enlarged and in 1938 a *Scuola Preparazione Capi* (Foremen's School) was opened.[13] Most of the students on this new course were those who had already done the three-year *Vedette* course.

One article in the factory newspaper in the 1920s tried to suggest that the *Vedette* school was primarily a philanthropic initiative, instituted to help the pupils, a result of the goodwill of the factory management: 'making the most of our young

[11] 'Gruppo Vedette Magneti Marelli', *SB* 6 (1927), 28. According to communist sources, however, the apprentices were not so well paid in this period as they earned no productivity bonuses: 'Dati e considerazioni sulle condizioni della classe operaia italiana. Le condizioni di lavoro degli operai della "Magneti Marelli" ', *Lo Stato Operaio* (15 June 1939).

[12] The first course had only 18 pupils but by 1927 there were 60 and this had risen to 120 by 1935.

[13] On this course the trainees had to do three hours of theory lessons after work hours for a year. After this, they were promoted to the higher grade of *operaio specializzato* and a few were picked to be foremen.

workers' potential . . . is such a high-minded, intelligent, and socially important task that it immediately attracted the active interest of our Management'.[14] Of course, the real motivation for the creation of this school was, in fact, hardly philanthropic; it was set up because of the difficulty of recruiting skilled workers. This problem was more acute at certain times than others. In 1935 with rearmament, for example, the need for skilled workers became very pressing:

The fact that work has been picking up again in industry, together with the introduction of the forty-hour week, has meant that the demand for labour has suddenly increased, especially in the engineering and electrical engineering sectors. In this factory too we have needed (and still need) to expand our labour force, but we are constantly confronted with the problem of finding enough properly trained workers . . . Well-trained workers are not easy to find.[15]

The shortage of skilled labour was not, however, confined to this period and, in addition to the recurrent problems that all engineering industries encountered in looking for sufficient skilled labour, this firm's production methods magnified the problem. It is true that in such a rationalized factory much work was being deskilled, so that many of the Magneti workers were probably actually doing jobs that required a lower level of skill than they might need elsewhere. But it is also true that rationalization meant that the firm needed a number of workers with new and different skills requiring special training. It was for this reason that the teaching provided by the newly created Fascist *Scuole Professionali* was deemed insufficient because too general and often inapplicable to the firm's specific needs.

Furthermore, the shortage of skilled workers was actually in part a result of rationalization itself, since previously skilled workers had learned through the artisan tradition:

in this way we provide the kind of practical teaching which has been so sorely lacking and the training is similar to that which young people could get in former times when they would start in a master craftsman's workshop as apprentices. At the beginning they would do the most humble type of jobs, such as clearing up the workspaces and machinery, keeping tools and components tidy, etc. Working alongside

[14] 'Gruppo Vedette Magneti Marelli', *SB* 6 (1927), 27.
[15] *SB* 6 (1935), 26.

the master craftsman they would slowly learn all of his skills working both manually and on machines, so that in certain cases the apprentice eventually knew the work better than the teacher.[16]

This same article from the factory newspaper also noted:

Those workers, as is well known, were a product of the artisan tradition, which the new organization of the work process has, apart from in a few rare cases, replaced everywhere.

In our opinion it is precisely this new organization of the work process which is the main cause of our current problems, since it is this which has produced, and will continue to produce, unless we do something about it, far too many of those workers who are known as 'mass-produced workers' in shop-floor jargon. This type of worker is frequently very bad at grasping productive innovations. If our work-force was composed entirely of this type of worker, it would be impossible to keep up with the constant changes and improvements in production. This is an extremely worrying problem for us.[17]

The school aimed to remedy these problems, yet it did not—as these extracts appear to suggest—simply mean that the management now provided the same type of training as that previously offered by an apprenticeship under a skilled artisan. At Magneti training was only to include that which was strictly relevant to the specific needs of production and was to include no 'unnecessary' information or theory. These workers were to become not generally skilled in a certain trade but trained particularly for this type of production and no other, learning no more and no less than was necessary to do the job properly.[18] This concept was fundamental from the time that the school was set up. In the 1920s, for example, during the sixteenth-month

[16] Ibid. 28.

[17] Ibid. 26.

[18] In this respect as in many others, the Magneti school was very similar to the one at Fiat: 'The cultural and theoretical foundation of the [Fiat] course, and especially the apprentices' course, was based on acquiring sufficient knowledge to understand the various phases of the production process without actually studying the scientific and technological background to this; in the two-year course the pupils learnt only the most basic theory, concentrating instead on practical work. Because of this, the pupils never acquired a broad understanding of the work process. In this way, they were living up to the ideals of [the Fiat engineer] Fossati who believed that workers should have *just sufficient education and no more*, to ensure that they could confidently carry out certain tasks without actually understanding the scientific theories behind the technological processes, and therefore, it was implied, without learning to question things which might lead them to go on to question society.' Sapelli, *Organizzazione*, 319–20.

course pupils learned little theory as their training focused on the actual production process of this factory. They had to become familiar with

> the different uses of the whole range of machinery which might be relevant to our production and especially automatic machines. There are also constant references to actual examples of shop-floor organization such as the system of the work cycle, timekeeping, etc. This course, which combines theory and practice, *is designed to avoid tiring the pupils with useless abstract notions* and instead to concentrate on getting them to understand everything they actually see and do, encouraging them eventually to prefer instinctively rational and scientific systems to empirical ones. This is especially true with regard to the boys on my course.[19] (my italics)

In 1935 the course was reorganized so that the *Vedette*, who were by now all male,

> study the entire cycle of work by doing a stint in each of the most important shops in the factory and the foremen in these shops help them learn while they are there . . . The *Vedette*, trained in this way, will be well prepared for our particular type of production.[20]

The workers that I interviewed who had been *Vedette* remembered it with affection. This school, for them, had been the starting-point of their careers. It was through this that many managed to become skilled and therefore better paid and more respected workers. As Dino said, to be a *Vedetta* 'meant to look forward. To look to the future, eh?' However, this type of training was also limiting for the workers, since it did mean that a *Vedette*-trained worker would not have the flexibility on the labour market of a worker trained under the old apprentice system, since much of the training was applicable in the context of this firm only. This same worker admitted:

> Because in the Magneti Marelli *Vedette* school, apart from teaching us a little maths, just the most important aspects of maths which were relevant to us of course, they used to teach us a great deal about our own machinery, so it was useless in another factory. (Dino)

As a result, many of the *Vedette* added to their already long and

[19] 'Gruppo Vedette Magneti Marelli', *SB* 6 (1927), 27.
[20] *SB* 6 (1935), 27.

tiring day by attending night school. As one ex-worker explained, after he had finished the *Vedette* course he enrolled at evening classes:

Because that was a school, the factory school, which was designed to meet the interests of the firm. They were interested in teaching young workers about the firm's own products. That's what we used to study, Magneti's own products. (Giglio)

The Foremen's Training Course was also a product of and moulded by the needs of production with similar strengths and weaknesses. In the Taylorized factory foremen had to know and understand a certain amount about rational organization, especially if they became head of a shop. These new foremen were provided with a more practical kind of training than the graduates of the engineering schools and also a more general understanding of production than a traditionally trained skilled worker, who concentrated on one trade only. The Foremen's School was designed to give workers who already had a certain amount of knowledge of production (they had usually already done the *Vedette* course or were skilled workers who had done their apprenticeships in other firms) precisely this kind of broad knowledge of scientific management. The course was also designed to inculcate them with a suitably deferential mentality to ensure that they would unquestioningly accept and interpret the orders of the new factory organizational hierarchy:

Our future foremen should have a very clear idea of the implications of scientific management and should know how to apply this in practice. In other words they should both have a spirit of initiative and also know how to obey orders, especially as regards respecting the pattern of the work cycle determined by the organizational departments.[21]

The workers' schools were, therefore, both a product of and essential to the Americanized organization of the work process. They were designed to produce a very different type of worker from the old highly skilled craftsman who had spent years of apprenticeship perfecting a trade. The new workers too were to be highly skilled, but the skills they needed and the type of instruction they received were strictly subordinated to the new needs of production. This meant that Magneti actually created

[21] *SB* 6 (1939), 41–2.

the type of workers they needed, since they could not be recruited elsewhere.

The schools aimed, furthermore, at creating a specific type of mentality. The more general lessons, for example, were not simply a product of the firm's desire to raise the general educational level of their work-force but were primarily moral and as much concerned with indoctrination as with education in the broader sense. They included not only work-related subjects like basic arithmetic, but additionally a good dose of political propaganda to cultivate patriotism and above all respect for discipline and the factory hierarchy.[22] Furthermore they helped reinforce gender divisions in the firm, since general education for girls meant simply domestic science whereas for boys it covered a wider range of subjects and in the build-up to war it even included 'pre-military' training such as marching and learning how to shoot.

The creation of the *Vedette* school and the other training courses was, therefore, an intrinsic part of the scientific reorganization of production. It tried to mould its pupils specifically into becoming the 'ideal' workers for a Fascist rationalized industry—knowledgeable in the new needs of production (but at the same time not too knowledgeable, for an overactive mind could also be dangerous to political stability), and also good Fascists with a healthy respect for their superiors at work.

'SCIENTIFIC PERSONNEL MANAGEMENT'

Another important element of Magneti's approach to 'human engineering' was the work of the Personnel Office and in particular its record system, which entailed keeping a personal file for each worker. One use of these records was undoubtedly to make it easy to weed out the unproductive or slow. But the factory newspaper preferred to dwell on the more positive

[22] With regard to the general classes on the *Vedette* course the factory newspaper wrote: 'Moral education is an integral part of this education and this aims to encourage a sense of duty and respect for the Nation' ('Scuola interna', *SB* 6 (1935)). The trainees in the Foremen's school, furthermore, studied technical and practical skills and also 'general lessons, Italian language, Fascist culture, and rules of discipline and behaviour' (*SB* 6 (1939), 41–2).

aspects and often stressed how the record system was used to enable the Personnel Office to decide which workers merited a rise in pay or a promotion:

> throughout his period of employment in the firm each worker is kept track of through the so called *foglio caratteristico*, on which all the relevant departments of the factory record details such as output, how hardworking the employee is, discipline, etc. At the end of every three months these files are examined and it is decided whether the worker merits a pay rise or promotion.[23]

This kind of record keeping was highly unusual and possibly unique in interwar Italy. It enabled the management to keep track of each individual worker at a central level and, just as the organization of production had been centralized, personnel management was not controlled by anonymous offices. Gone were the days when each foreman on the shop-floor knew what sort of recruits were needed, chose them on personal judgement, and watched over who was a good worker and who was lazy or undisciplined. Now each worker was recruited according to the labour requirements drawn up on the schedules for each product and selected according to so-called 'scientific' criteria. From the day of recruitment a careful centralized record was kept of aspects such as discipline and productivity. Rates of pay and the pace of work were established centrally and the foreman's role was reduced to pointing out which individuals should have a small extra pay rise.

INDUSTRIAL PSYCHOLOGY[24]

Complementing the work of the Personnel Office was the Industrial Psychology Unit. As Fiat engineer Mario Fossati

[23] 'Le possibilità pratiche di organizzazione scientifica del lavoro in Italia. (Il fattore umano)', *SB* 5 (1927), 21–2. This recruitment system was also noted by a clandestine Communist Party report in 1936: 'A filing system is used. When a new worker is taken on a file is drawn up containing details of his work record etc. This is given to the head of shop every 3 months, so that it can be decided if the worker should be given a rise of 5–10 centesimi per hour' ('Lettera da Benvenuto', 3 June 1936, APCI, f. 1376, fo. 42–3).

[24] On industrial health and industrial psychology in this firm see P. Willson, ' "The Golden Factory": Industrial Health and Scientific Management in an Italian Light Engineering Firm. The Magneti Marelli in the Fascist Period', in P. Weindling, *The Social History of Occupational Health* (London, 1985).

noted in 1927, at Magneti new workers were, 'for the first time in Italy, scientifically recruited, which means that a worker is only taken on if he fulfils certain requirements, and he is assigned to the shop to which he is most suited both mentally and physically'.[25] Run by the factory doctor, Annibale Correggiari, this unit developed a supposedly 'rational' procedure for selecting new employees. This consisted of a series of tests which each prospective employee had to undergo. Specialized equipment was used, such as the Beyne and Bettagne apparatus, which measured the speed of reaction to a visual stimulus. The worker tested had to interrupt an electric current as soon as a light became visible. A mechanical pen automatically recorded the speed. Other equipment aimed to measure such factors as the level of tiredness experienced by a worker at different stages of the day, the sensitivity to touch of the fingers, physical strength, or accuracy in drawing a line exactly half-way between two others.[26]

Workers were tested in this way principally to raise output by improving the 'quality' of the workforce. The ideal score for each of these tasks was determined by testing existing workers. However, this recruitment strategy aroused the suspicions of the workers, who feared that it could be used to eliminate the weak, ill, or ageing who often most desperately needed work. Umberto Quintavalle specifically tried to refute these allegations in an article in 1928. He argued that the industrial psychology test was not actually used for this purpose, but, instead, ensured only that workers were not given jobs too taxing for their capacities.[27] But evidence from the factory archive suggests that the workers' fears were not unfounded. During the crisis of the early 1930s the unit could have its uses in cost cutting.[28]

In addition, delving a little deeper under the management

[25] M. Fossati, 'La selezione del personale in una azienda industriale di grande importanza' (prima parte), *SB* 6 (1927), 24.

[26] Fossati, 'La selezione del personale' (seconda parte), *SB* 7 (1927), 21–2.

[27] U. Quintavalle, 'L'industria e la lotta contro la tubercolosi', *L'assistenza sociale nell'industria* (ASNI), (1928), 8.

[28] e.g. in 1933 a memorandum to Bruno Antonio Quintavalle concerning the sacking of a worker noted that 'At the moment the management is reviewing the workforce to eliminate those employees who are the least productive because of their age or temperament, and replacing those dismissed with the young and energetic' (letter with illegible signature, 24 Mar. 1933, APMM).

propaganda about its 'impartial' and 'scientific' approach, it is worth noting that, in reality, science had by no means made obselete the time-worn system of personal recommendations. Many aspiring workers armed themselves with reference letters from local Fascist branches or parish priests,[29] and the men and women that I interviewed confirmed that, in order to obtain work at Magneti, more than simply luck, a willingness to work, and a high score in the industrial psychology test was necessary. In fact, a number of the interviewees had obtained work through knowing someone in the factory. Some used recommendations from official persons such as the mayor of Sesto, but often it was simply a relative or a friend already in the firm. One woman even got her job because she was the girlfriend of a head of shop. Another ex-worker recalled how his mother was able to use her position as a doorkeeper at Magneti to obtain work for a large number of people, including numerous members of her own family, 'my mother had the possibility. She had, as I was saying, she had the power to recruit people' (Ercole). Sometimes additional (and hardly scientific) persuasion was needed, as Maria V. recalled:

> my father knew the doorkeeper at Magneti because he lived in our village . . . So—I'll always remember it, oh my goodness!—he gave me (I'm ashamed to say) a letter to take to him, and they said to me: 'Yes, yes come tomorrow morning'.—So the next day my father made up a basket with a chicken inside it to take to that man (because they'd given me the job at Magneti!) . . . Oh Lord, I was in the train with all that stuff . . . imagine a girl going around with a chicken in a basket and everyone was thinking '. . . goodness knows where she's going!'

INDUSTRIAL HEALTH AND SAFETY

Apart from recruitment, the Industrial Psychology Unit was also involved in the firm's health and safety policy where it worked alongside an American-style safety committee. In this field too Magneti had an unusual approach.

Right from the early period of industrialization there had been some interest in the problem of industrial health in Italy, giving

[29] Many files in the factory personnel archive contain such letters.

rise, for example, to the introduction of some rather ineffective legislation[30] to make insurance against certain industrial diseases compulsory in January 1904. More spectacularly in 1910 the world's first industrial health clinic was opened in Milan by Luigi Devoto. To a certain extent early industrial health legislation was inspired by fear of the strength of the workers' movement around the turn of the century. With the arrival of the Fascists, however, the interest in health and safety did not totally disappear and, in 1934, compulsory insurance was extended to cover six more industrial diseases.[31] 1933 saw the creation of INFAIL (Istituto nazionale fascista per l'assicurazione contro gli infortuni sul lavoro), a state organization which centralized all the insurance bodies.[32]

The rise to power of Mussolini's party did, however, create problems for some members of the medical profession, since the destruction of the free trade unions undermined the relationship of co-operation which had begun to evolve between certain factory doctors and workers in the pre-Fascist period. In spite of this, the inter-war period was not entirely negative in the field of industrial health. As Luisa Dodi Osnaghi has argued, some members of the medical profession were becoming increasingly interested in this question.[33] Certain medical experts were beginning to look beyond the narrow definition of industrial health, which simply saw it in terms of dangers in the workplace. Factory doctors, they stressed, should take a broader view of workers' health.

Despite these advances in medical awareness of occupational

[30] In 1906 a circular from the Ministry of Agriculture, Industry and Trade noted that the law was 'generally respected very little. When factory inspectors complain, the majority of industrialists and entrepreneurs reply that they have never even heard of the legislation' (R. Romano, 'Gli industriali e la prevenzione degli infortuni sul lavoro (1894–1914)', in M. L. Betri and A. Gigli Marchetti (eds.), *Salute e classi lavoratrici in Italia dall'Unità al fascismo* (Milan, 1982), 143).

[31] Lead, mercury, phosphorus, carbon disulphide, and benzol poisoning, and ankylosis.

[32] On social insurance legislation see A. Cherubini, *Storia della previdenza sociale in Italia 1860–1890* (Rome, 1977); G. Bronzini, 'Legislazione social ed istituzioni corporative', *Annali Feltrinelli* (1979–80).

[33] Much of this interest and research was focused on the problem of toxicity, as the chemical industry was expanding and new processes and materials were being introduced (L. Dodi Osnaghi, 'Aspetti della condizione operai e della nocività attraverso le riviste di medicina del lavoro', in *Annali Feltrinelli* (1979–80), 234).

health questions, conditions in many industries were very bad. In sectors such as textiles this was frequently made much worse by the particular way in which 'rational' management methods were introduced into Italian industry with an emphasis on speed-up. 'Rationalization' implied a whole new range of industrial health problems, but instead of looking for new solutions, contemporary analysis tended to blame the victims. In the opinion of many occupational health experts,

> with increasing frequency workers were inflicting injuries on themselves in order to obtain whatever small compensation the law provided . . . this accusation that injuries were self-inflicted was the subject of a propaganda campaign during the whole [inter-war] period which became even more insistent during the debate about the reform of the 1904 legislation.[34]

The employers' organization, the *Confindustria*, for example, claimed that only 8 to 30 per cent of all accidents were caused by machinery, whereas 70 to 75 per cent could be attributed to the workers themselves.[35] The logical conclusion of this was that if workers caused accidents they should be sacked. Sapelli argues that many employers did not bother to introduce safety measures, since no unions safeguarded workers' interests. He sums up the policy of the majority of firms as the 'selection and elimination of unhealthy workers and their replacement with younger elements whose health had not yet been impaired by factory work'.[36] If this can be taken as a fair evaluation of health and safety standards of most firms in this period then Magneti must be seen as a somewhat exceptional case.

The variety and complexity of production at Magneti created a huge range of potential health hazards. For example, products such as radios and car parts required dangerous shearing machinery; batteries involved toxic chemicals; the foundry plant presented numerous risks with its furnaces and sandblasting equipment. Despite this, the evidence that has survived[37] of accidents and health problems is largely restricted to workers in

[34] B. Bottiglieri, 'Razionalizzazione del lavoro e salute operaia tra le due guerre: l'atteggiamento del sindacato e del governo', in Betri and Gigli Marchetti, *Salute e classi lavoratrici*, 871.

[35] Sapelli, *Organizzazione*, 375.

[36] Ibid. 372.

[37] Various letters and memoranda concerning industrial health are to be found in the APMM.

the Shearing Shop, who frequently injured their hands, and those who contracted silicosis from handling chemicals or working on the sandblasting equipment. The fact that accidents do not seem to have occurred so frequently in other jobs suggests that the health and safety policy may have had some effect.

In June 1927 Magneti Marelli became the first factory in Italy to set up a Safety Committee. Sapelli argues that, whereas these committees fought the industrial accident rate seriously in their country of origin, the USA, in Italy their role was very limited, concentrating on simply trying to teach workers to be more careful and watching out for those who were unsuitable for their jobs in order to dismiss them.[38] This may have been true in other industries but it would be an over-simplistic assessment of the Magneti Marelli Safety Committee.

This committee consisted of a group of foremen, workers, and technical and managerial staff. The factory doctor, the social worker, and other representatives of the firm's welfare services were members, although by the 1930s their role seems to have diminished to an advisory one. The committee worked in co-operation with all sections of the factory including the Organizational Department. The exact composition and structure of this committee was modified on a number of occasions but two important features remained constant. One was that, although this was ostensibly a mixed group of workers and managers, the workers remained essentially subordinate to the management. The other point is that, despite the large numbers of women employed in this firm, the Safety Committee was composed entirely of men, with the sole exception of the factory social worker. Even though many accidents occurred in female-designated jobs, such as shearing, there is no evidence that a single female member of the work-force participated. Despite appearances, the Safety Committee was not a democratic power-sharing exercise. Yet the workers were not included simply for window dressing. To understand this, it is necessary to look at how the Safety Committee worked.

When accidents occurred, the committee investigated. Sometimes this simply meant collecting a report from the foreman,

[38] Sapelli, *Organizzazione*.

but frequently it entailed carrying out a detailed investigation or even getting the help of the worker involved to reconstruct the event. In this way, the committee gathered statistics on the seriousness and frequency of accidents, as well as investigating their causes. Recommendations were usually made on the prevention of future accidents, such as the introduction of protective clothing or shields. In drawing up the report, the committee worked in co-operation with various technical and managerial departments of the factory.

For example, when Adelina, a worker in the Magneti-making Shop, cut her finger on a grinder, the Safety Committee made a thorough investigation. Not only was this particular accident taken into consideration, but also the records of previous similar cases were examined in the search for a solution. The report concludes that another similar accident could not be prevented in the future simply by installing better protective equipment. The matter was, therefore, referred to the Organizational Department, which proposed three solutions. One of these suggestions has, unfortunately, been lost and the second was simply a mechanical adaption, but, significantly, the third proposed an actual alteration of the cycle of the work process, so that the order in which certain tasks were carried out was modified.[39] If this case is typical, it would suggest that the firm had a broad and innovative approach to accident prevention.

The committee was also involved in an active educational campaign. The workers who were members were important in this, since they were drawn from every shop of the factory and could, therefore, relay new ideas on health and safety to their work companions. A special notice-board was set up, with photographs, stories, and advice, featuring slogans such as 'A pay packet is always bigger than injury benefit' or 'A word of advice to a workmate costs nothing and can save him from an accident'.[40] This also presented statistics and information about accidents that had already happened and accounts of the Safety Committee's activities. To encourage interest, humorous stories were invented about a fictitious character called *Infortunello* (derived from the word for accident, *infortunio*) who had every

[39] Unsigned memorandum [1935?], APMM, f. 'Adelina S'.
[40] 'In casa nostra', *SB* 10 (1929), 51.

conceivable form of accident. There were also competitions to find the work team with the lowest accident rate. Safety booklets were produced for the work-force and visits were organized to exhibitions. Numerous articles were published in the factory newspaper.

One of the aims of this campaign was, undoubtedly, to stress how much the firm was doing for the work-force. The articles in the factory newspaper, for example, were generally focused on the *active* role of the firm in fighting the accident rate and industrial disease. Less attention was paid to the dangers presented by particular types of machinery, except in the context of describing new protective innovations. The management was clearly most particularly interested in convincing the workers that their welfare was adequately cared for. The campaign also tried to encourage the workers to participate personally in the fight against industrial accidents by taking greater care during their work, suggesting new ideas for accident prevention, or serving on the Safety Committee. Literacy rates were high among the work-force, so that the majority of them had access to the educational campaign's materials, but it is difficult to assess whether it had any actual effect, although there is some evidence that workers did make suggestions about making machinery safer.[41]

The Magneti Marelli Safety Committee saw industrial health as a complex issue, involving a mixture of technical, organizational, environmental, and human factors. The last two were also the concern of the Industrial Psychology Department. The management liked to stress the importance of the Industrial Psychology Unit in the fight against industrial disease, probably to allay the fears of the work-force about its other role. The testing procedure, it was argued, was important in the health and safety campaign in that it supposedly only recruited workers who were suitable for the job and therefore not 'accident prone'. In other words the worker was to be fitted to the needs of the machinery. On the other hand, the unit was also concerned with the environmental causes of accidents and disease. Light, heat, and ventilation were all monitored and attempts were made to improve the position in which each

[41] Both articles in *SB* and notes in APMM mention workers who have suggested safety modifications.

worker had to sit or stand. For the measurement of light, for example, a special machine, the 'Luxmetro', was used. A table was drawn up of the number of 'Lux' units necessary for each task and the machine was carried through the factory making checks.[42]

The work of the Safety Committee and the Industrial Psychology Unit shows that Magneti Marelli took the question of industrial health seriously compared with many firms in this period. It is not easy, however, to evaluate to what extent this campaign was actually successful. The only figures that I have been able to uncover are the firm's own statistics showing the effects of the accident prevention policy in the first two years after the foundation of the Safety Committee. According to these, published in an article by Correggiari, the number of accidents in 1928 was only 76 per cent of the 1926 level.[43] There was no drop, however, in the amount of production time lost in short-term sick leave following accidents. Correggiari dismisses this, claiming it was most probably simply due to the fact that some of the staff of the medical centre were more generous than others when they granted sick leave.[44] An alternative explanation might be, however, that the seriousness of smaller accidents was increasing so that longer periods of time were necessary for recovery. What did drop was the firm's expenditure in permanent disability compensation. There was a clear reduction in the seriousness of accidents which caused permanent disability, and since the compensation for this represented a much bigger expense than the loss of working days in temporary sick leave, Correggiari was able to conclude that the overall result of the accident prevention policy was an impressive saving for the firm. By 1928 the cost of industrial accidents per 1,000 hours worked had fallen to 70 per cent of the 1926 level.

These figures are clearly too limited to serve as a yardstick by which to judge the success of the policy. If they are accurate, then they do suggest some positive results at least in the initial years of its existence. Unfortunately, I have not been able to find any comparable figures for the 1930s, nor any indication of whether industrial diseases became less common.

[42] 'Per difendere gli operai', *SB* 6 (1929), 52.
[43] A. Correggiari, 'La prova dei fatti', *SB* 1 (1930), 45.
[44] Ibid. 46.

Whether successful or not, these initiatives add up to an innovative health and safety campaign in comparison to many other contemporary firms, and they testify to the fact that the Magneti Marelli management and the factory doctor Correggiari were prepared to experiment with American ideas that had not yet been tried in Italy. The factory archive does, however, reveal another side to this apparently harmonious picture. The firm was keen to safeguard the work-force's health, but the question of silicosis demonstrates that they were not willing to accept any questioning of their ability to do so. The cases of silicosis also suggest that the health and safety policy may have altered with time and the earlier seriousness may have given way to a more strong-arm approach. It is, however, very difficult to be certain about this, since the absence of documents relating to the earlier period may simply mean that they have been lost.[45]

In Italy silicosis had traditionally been seen as a miners' disease; it was only slowly recognized that it could also be an occupational health risk in certain engineering jobs. Since the Italian mining sector was small, it was considered rare and the 1934 legislation which made compensation automatic for six industrial diseases omitted silicosis from the list.[46] In spite of this, a number of Magneti Marelli workers did attempt to get compensation in the late 1930s and early 1940s.

Most of these sued after the firm sacked them for reasons of ill-health. Only one record survives of a worker actually being paid compensation and the sum was fairly small—equivalent to about eighty days' pay.[47] The firm won certain cases by arguing that the problems that certain workers had with their lungs were due to other causes, such as bronchitis, whose origin was outside the workplace. Yet medical journals in the 1930s show that the link between silicosis and sandblasting, or any other job carried out near it, was understood. Many contemporary medical experts were aware of these dangers, and, furthermore, they considered that the use of ventilators and sealed-in cabins such as those cited by Magneti in its defence, was insufficient protection.[48] The firm was willing to pay high fees for lawyers

[45] This is because the materials contained in the APMM are generally much richer for the 1930s than for the preceding decade.

[46] Dodi Osnaghi, 'Aspetti', 274 n.

[47] APMM, f. 'Aristide I'.

[48] Dodi Osnaghi, 'Aspetti', 277.

and professional experts, which suggests that it was less concerned about the immediate expense of paying compensation to the particular worker concerned, than about the danger of setting precedents. This may mean that the eight surviving records of cases of workers who took the firm to court were only a fraction of the potential number. The case of Francesco demonstrates the atmosphere in which workers tried to get compensation.[49]

Francesco worked, in 1935, in the Heat Treatment Shop, next to a vat of potassium cyanide. Fumes from the vat permanently damaged his lungs, as was recognized by the factory doctor (by now a certain Dott. Fulghieri), who recommended that he should be transferred on to other work. But Francesco wanted compensation since his health was ruined, and he took the firm to court. He was unable to get compensation, however, until after the fall of Fascism and the Liberation, when the firm was temporarily run by Brasca, the managing director appointed by the CLN. In a letter to Brasca, in 1946, Francesco described how the personnel manager of the firm tried to prevent the court case by promising a friendly settlement.[50] No money ever appeared, and, as he continued to press for justice, the promises of cash payment turned to threats until he had to drop the matter. By this time, it was 1942 and workers who caused trouble could be mobilized into the army.

Apart from the cases of silicosis, however, the firm's policy was impressive compared with most contemporary firms. Certain articles in the factory newspaper tried to argue that the firm was principally interested in health and safety for patriotic motives. But this rhetoric was merely yet another example of the firm trying to make political capital out of a pre-existing policy.[51]

[49] APMM, f. 'Francesco G'.

[50] Letter to Brasca, signed Francesco G., 12 Dec. 1946. (ibid.).

[51] One article in the factory newspaper, written by one of the more active Fascists amongst the firm's engineers, even maintained that the interest in preventing industrial accidents was principally motivated by the Fascist demographic campaign, since injured workers could not fight as soldiers: 'Amongst the permanently injured the level of handicap varies, but the statistics suggest that at least a third of these would no longer be capable of serving actively as troops in defence of the Motherland; when added to the number of deaths this amounts to an annual loss of about 18,000 men, the size of an entire Division, which is far from negligible' (Vico d'Incerti, 'La lotta contra gli infortuni', *SB* 4 (1938), 37).

The interest in health and safety was more than just an attempt to give the firm a reputation as truly Fascist. Neither was it purely an old-style paternalist policy, since this firm saw industrial health and scientific management as closely interrelated.

Magneti Marelli, strongly influenced by American ideas, saw health and safety as more than simply a question of protecting the workers' well-being. Correggiari stressed this, in 1929, when he claimed that, in this firm 'accident prevention is based on the principle of *rational organization*. In the eyes of both the management and the workers, the link between scientific management and accidents is perfectly clear' (his italics).[52] He does not, however, clarify exactly what he means by this, and it is worth looking more closely at the nature of this link.

To a certain extent, it is true that a dynamic health and safety policy was actually necessitated by the introduction of American-style production methods which created new and unprecedented hazards. Production tempos were much faster and at this accelerated pace workers could be under constant stress and therefore at risk. Umberto Quintavalle recognized this when he wrote, in 1928, that

> the faster work tempo imposed by modern rationally organized production only worsens the [health and safety] situation, since the nervous tension and increased physical effort required undermines the workers' resistance, unless we also introduce preventive measures.[53]

Not only the pace but also the type of work could cause problems as, to a great extent, scientific management meant the division of work into essentially boring tasks which dulled the workers' senses and made them less careful. There were further dangers inherent in the rapid technological innovation; constantly changing machinery and new processes meant unknown health risks. In contrast to some other firms who tried to deny the existence of risks specifically created by the new approach to management, all these problems were taken seriously.

Furthermore, the 'scientific approach to management' enabled the firm to see the industrial health policy not as a cost and burden, but rather as a self-financing measure. At first this was

[52] A. Correggiari, 'Esperimenti antifortunistici', *SB* 1 (1929), 25.
[53] Quintavalle, 'L'industria e la lotta', 8.

not mentioned and an article in the factory newspaper claimed that workers' health was simply a question of the benevolence of the employers.[54] However, this attitude was soon abandoned and various articles on the health and safety policy emphasized that it paid for itself, since accidents which disrupted the flow of work were costly, sick leave and compensation meant lost production time and expense, and healthy workers were more productive. In fact, Correggiari himself admitted that the management was not willing to spend any more than they calculated would be returned in this way:

It may seem that these provisions are too expensive and sometimes have an unfavourable effect on productivity: as far as cost is concerned, it should, of course, always be kept in proportion to the increased profit yielded, otherwise it would be uneconomic.[55]

He did, however, maintain that there was rarely a conflict between the needs of production and the needs of health, and strict budgeting did not lead the firm to scrimp on protective measures. Furthermore, in the same article, he went on to argue that greater safety protection could actually raise productivity and overall profits, since

dependable accident prevention measures eventually have a psychologically favourable effect on the worker: relieved of the danger which previously threatened him he can dedicate himself more calmly to this task, feeling himself in safety he speeds up his pace of work and sometimes *solely due to the safety measures*, there is a net increase in production.[56] (his italics)

But what the management really liked to stress was the fact that although scientific management caused health risks it also, conversely, could facilitate an innovative approach to the problem. The tools of scientific management could themselves be used in the fight against industrial disease and accidents. Thus, it was claimed that the health and safety policy was, in itself, 'rationalized'. The work of the Industrial Psychology Unit is a very clear example of this. Its work in filtering out unfit workers was a clear source of suspicion for the workers, and in order to win a consensus for this type of Americanized

[54] 'Assistenza sociale', *SB* 7 (1925), 11.
[55] Correggiari, 'La prova dei fatti', 46.
[56] Ibid. 47.

innovation it was important to present technological and managerial innovations as beneficial for all.

'RATIONALIZING THE HUMAN FACTOR'

Through an interest in 'the human factor', the management argued, the firm had got away from mere charitable welfare. They had found, for example, a way of helping the workers by protecting their health and safety, endangered by the new methods of work, at no cost to the firm. As productivity rose, firm and work-force would work together in happy co-operation to everyone's mutual benefit. It is easy to see how these ideas fit in admirably with the Fascist mirage of class co-operation and also with the idea that everyone could profit mutually from the introduction of scientific management.

The initiatives described in this chapter were both an integral part of Magneti's concept of scientific management and important in the firm's attempt to create a consensus for it. They demonstrate that the management took a broad view of Americanization which contrasted with many other firms in this period.

Nevertheless, the 'benefits' the workers derived from these initiatives remained strictly within the management's control and definition. In the workers' school the apprentices gained a valuable training but were never allowed to get enough theoretical grasp of the overall nature of the productive process to challenge management authority. General education for girls was confined largely to domestic science. The cases of silicosis reveal the essentially authoritarian nature of these initiatives, for the management may have taken the issue of health and safety seriously but they would not allow their competence in this matter to be questioned by the work-force. Although the Safety Committee did contain workers, they remained subordinate to the managerial representatives. In the end it was the management alone who reserved for themselves the exclusive right to define what constituted suitable education, health, or ill-health for the work-force.

The American methods constituted a hierarchical and essentially undemocratic system of management and, as the cases of silicosis described above demonstrate, the co-operation policy operated against a background of threats and intimidation. If the workers found cause to doubt the effectiveness of the firm's provisions, they had little chance of making their voices heard, a fact which is abundantly reinforced by a consideration of the pay systems which were introduced.

4

'Taylorismo' at Magneti III: Beating the Clock

WAGE incentive systems were an integral part of any approach to 'scientific management' in this period and according to some American advocates of such methods high wages should and could be the result of the reorganization of production. Henry Ford himself called for a high-wage policy and criticized wage reductions as leading to 'the cutting of buying power and the curtailing of the home market'.[1] He also asserted that 'it is not the employer who pays the wages. He only handles the money. It is the product that pays the wages and it is the management that arranges the production so that the product may pay the wages.'[2] This suggestion that wages were in some sense 'automatically determined' was of course nonsense. Increased productivity and better wages did not necessarily accompany one another. In Italy it was easy for employers to keep wages low, given the destruction of the free trade unions, the outlawing of strikes, and the virtual powerlessness of the Fascist unions in the fixing of base and bonus rates. In many firms in the 1930s, furthermore, the particularly exploitative bonus system invented by Charles Bedaux[3] was introduced.

Charles Bedaux claimed to have discovered scientific methods of measuring the precise relationship between work and fatigue and that it was possible to determine the exact proportions of rest and work required for any given task. The Bedaux experts would calculate the number of 'B' units (sixty to an hour) required for any given task. There is, however, no evidence that

[1] H. Ford (in collaboration with S. Crowther), *My Life and My Work* (London, 1924, first published 1922), 116.

[2] Ibid. 78.

[3] See C. E. Bedaux, *The Bedaux Efficiency Course for Industrial Application* (Grand Rapids, Michigan, 1917); H. S. Person, 'The Bedaux System', *New Republic*, 93 (24 Nov. 1937).

Bedaux actually did any scientific experiments to calculate these rest–work relationships, and E. Layton has argued that basically he was little more than 'an opportunist and his system . . . a perversion of scientific management'.[4] Yet Bedaux had not himself really introduced anything that was fundamentally different from Taylor's ideas. What he did offer that was new, as Duccio Bigazzi has aptly commented, was 'a marketable form of scientific management methods which he sold carefully packaged to suit the varying "tastes" of his clients in different countries'.[5] More than any other incentive system used in this period the Bedaux provoked discontent and unrest wherever it was introduced. This was partly because an introduction of this system raised output far more than wages, and partly because it was so complicated to calculate that workers often did not understand the way in which their wages had been worked out and this led to distrust and confusion.[6]

PAY SYSTEMS AT MAGNETI

The workers in Italy's most 'scientifically' organized factory were not all paid in a uniform manner. Most had their earnings calculated according to a bonus incentive system, but throughout the inter-war period a section of the Magneti work-force continued to receive a fixed hourly rate. These ranged from certain unskilled workers such as guards and cleaners to the very highly skilled fitters, the common denominator which united them being the fact that their pay could only with great

[4] E. Layton, 'The Diffusion of Scientific Management and Mass Production from the U.S. in the Twentieth Century', *Proceedings of the 16th International Congress in the History of Science* (Tokyo, 1974), 382.

[5] D. Bigazzi, 'Organizzazione del lavoro e razionalizzazione nella crisi del fascismo 1942–3', *Studi storici*, 2 (1978), 269.

[6] In Britain e.g. 'in 1932, the Trades Union Congress instituted an enquiry into its operation, which revealed that such unions as had had experience of it in their trades were in nearly every case hostile to it. A well-known authority on labour management declared that it yielded lower hourly earnings than most other American bonus systems. Its general effect apparently was to stimulate an increase of output that was proportionately much greater than the increase of pay that it yielded. Enquiries in America showed that it resulted in increases of production of about 50% on the average and increases of earnings of only about 20%.' (M. Dobb, *Wages* (Cambridge, 1960; 1st edn. 1928), 64).

difficulty be linked to productivity by means of a bonus scheme.

Clerical workers too were paid according to a flat rate, but in the form of a monthly salary. In some firms in this period, timekeeping was being applied to office work,[7] but at Magneti there was no attempt to subject any clerks to this, even those engaged in monotonous, repetitive tasks such as card-punching. Admittedly productivity incentives were built into the calculation of their pay in the shape of occasional rises and end-of-year bonuses, but this was, of course, a far cry from real Taylorized incentive systems. In reality, the clerical workers who did the work of implementing the bonus scheme were protected from it themselves. The existence of a different pay system for clerks helped reinforce the pecking order of the factory whereby white collar (and indeed 'white blouse') workers enjoyed far higher status than those on the shop-floor and served to cushion the potential resentment of deskilled clerical workers by emphasizing the idea that 'rational' innovations in clerical work were different from those affecting manual workers.

On the shop-floor itself most of the work-force were paid by a bonus scheme. In some cases this meant an individual bonus, whilst others (probably the majority) were paid collectively according to the output of their work team. Usually all the members of the team, even including relatively highly paid employees such as heads of shop, had their wages calculated according to the bonus the team earned.[8]

It has proved difficult to be get more than fragmentary information on the way in which pay rates were calculated at Magneti. The factory newspaper, which talks copiously about planning departments and quality control, is rather more reticent on the subject of which pay system was actually introduced and it has not been possible to establish exactly which method was used in the early period. What is fairly clear

[7] G. Sapelli, *Organizzazione, lavoro e innovazione industriale nell'Italia tra le due guerre* (Turin, 1978), 207–26.

[8] 'Lettera da Benvenuto', 3 June 1936, APCI, f. 1376, fo. 42. In this letter Magneti is referred to as Factory 'C'. The collective bonus system does not seem to have been common in Italian industry in this period, although it was introduced at Fiat's Lingotto plant between 1921 and 1927. In contrast to Magneti, at Fiat the bonus was calculated collectively for the whole plant rather than for each team. See D. Bigazzi, 'Gli operai della catena di montaggio: la Fiat 1922–1943', *Annali Feltrinelli* (1979–80), 903.

is that, by the late 1930s at least, the Bedaux system itself was being used. According to one ex-foreman who worked at Magneti in the 1930s, the Bedaux system of calculating piece rates was introduced in 1936:

In about 1936 they brought in an incentive systems expert to Magneti and they wanted to introduce the famous Bedaux system—you know Bedaux like Taylor—which meant that the work had to be done in a certain amount of time which was calculated and measured by special technical staff. For example, while a certain part was being made they would stand there with a chronometer in their hands and they would time a worker. They would then work out the bonus rates according to the amount of time the worker had taken. [Q. Didn't they have the Bedaux system before then?] Before then, no. There was a bonus system but it wasn't calculated in that way which, according to them, was more scientific. (Santino)[9]

Santino recalled that previous to this times were calculated simply by the foremen according to their personal judgement. But this picture of a rather rough and ready system in the early period is perhaps somewhat misleading. This may have been so in the shop where he worked, but in others it is clear that even before the introduction of the Bedaux itself Magneti was making use of some sort of 'scientific' bonus system to stimulate productivity. From the 1920s an Elementary Timekeeping Office (*Ufficio Tempi*) existed and the shop floor already seems to have been overrun by a veritable army of timekeepers. According to *Sprazzi e Bagliori*, by 1928 the firm employed detailed 'Taylorist' timing in the Magneto Shop.[10]

Even in this early period the complexity of the operation of the bonus system seems to have necessitated a great deal of clerical work, and this was further exacerbated by the rapidly changing nature of production. Each worker had to be timed, rather than just a few from each shop, and retiming had to occur constantly as new processes and products were introduced. One of the pay clerks recounted how special books were used to calculate the rates:

[9] Another testimony by Santino, who was a member of the clandestine Communist Party with the code name 'the poet', appears in a book edited by Franco Alasia. See 'Il poeta' in F. Alasia (ed.), *La vita di prima* (Milan, 1984).

[10] *SB* 6 (1928), 25.

These books which defined the standard times for each process were made up gradually as more and more of the production was included. The timekeepers went into the workshops, recorded the times, doubled the actual times compared with the times allowed, and then made the calculation according to the category of the worker. (Maria B.)

. . . they were practically registers. Registers where you could look things up . . . yes it was complicated, because every operation was given a symbol, a code . . . it was an A, a B, a C, and a D and then they were catalogued . . . according to the job and what purpose it served. It was all written in these books. At first in books, then they transferred them to cards. That made it easier to find things than using books. (Maria B.)

By the late 1930s the work of continual retiming seems to have proliferated so much that attempts were made to apply mass-production principles to the process of timekeeping itself. An article published in *Sprazzi e Bagliori* in 1939 went to great pains to stress how 'scientific' and 'rational' this new method was:

We have now definitively started using the most economical method of calculating times, which consists in building up a list of standard rates which can be used wherever that particular operation is carried out throughout the factory. This will eventually save us a great deal of time study work . . . It is very important that we follow the rules laid down in the collective contracts, according to which the bonus rates must remain stable unless there are any changes in the productive process or the type of tools or machinery used. Because of this we have to take care to keep very detailed records of how the times are calculated so that, if necessary, we can justify any changes, which in fact happens frequently since our production methods are constantly being updated in order to minimize the cost of each product. With the standard times method which we referred to above, we can, in most cases, alter the bonus rates without the need to re-time each process on the shopfloor.[11]

The author of this article is keen to emphasize the complexity of the operations involved:

The work of timing carried out on the shopfloor is, in fact, only one part of the calculation of standard times, although it is probably the most laborious part. The calculation of times by the subdivision of each operation into basic parts means recording the relevant times,

[11] E. Loi, 'I tempi di lavorazione sui torni Pittler', *SB* 4 (1939), 9.

evaluating the output of the particular worker recorded, reducing the observed times into standard times (i.e. the time needed by an average worker as specified in the collective contracts), comparing different times taken by various different workers to avoid subjective judgements, defining and tabulating the various ways in which the standard times can differ according to the various sizes of components produced, studying of the cutting times of various tools, defining standard rest periods, etc.[12]

Such articles aimed to convince the work-force of the justness and impartiality of management methods which could 'scientifically' determine what constituted a fair reward for a day's work. If this were true then, as Taylor had argued, unions and collective bargaining should no longer be necessary as conflicts would not arise. However, in spite of the firm's attempts to persuade them, the Magneti workers appear to have been less than totally enthusiastic about the virtues of Taylorist science and the ability of the management to be 'objective' about their pay.

WORKERS AND THE BONUS SYSTEM

In 1926 the communist newspaper *Unità* reported that faced with the firm's battery of 'scientific methods', many of the workers could not understand how their pay had been worked out:

> Let's talk about the bonus system now. This has two forms—a fixed percentage and the incentive system. Only a few workers get the fixed percentage and they are the lucky ones because at least they can work out how much they are supposed to be getting. Those who are earning according to the incentive bonus can never, absolutely never, work out how much they should earn. Furthermore, sometimes certain teams don't get any bonuses at all for months on end because the factory says that they are making a loss. The bonus system in this firm is in a state of chaos. No one can understand it at all. If a worker wants to know something about it he won't get a reply. This confusion is due to the fact that the foremen get the bonus with the workers and this reduces the workers' earnings even more because the foremen have high wages per hour and work longer hours. The bonus rates for each part are

[12] Ibid. 10.

known only to the foremen and they are kept hidden from the workers.[13]

Oral testimonies suggest that this problem persisted and was still an issue in the late 1930s. As one man commented:

No, no. It wasn't possible to work out [how the bonus was calculated]. You just took your paypacket, you read on it that you had so much in bonus. It wasn't much, it really was very little. You just took it . . . You couldn't check if it was right. There was no way of checking. (Pierino)

It appears to have been especially hard to understand how fairly the rates had been calculated for those with a team bonus:

I don't know if you can imagine. Ten for you, ten for this one. The same for everyone. And you couldn't tell how many parts you yourself had made, how much time you'd taken, what your own rate should be, because one worker would have one pay level, another got a bit less, and so on. The whole thing was a bit like that and it was pretty hard to check if the bonus was right. (Pierino)

Problems with the pay system were particularly acute in the period of economic depression. Initially in the *Quota Novanta* slump of the late 1920s this was not so severe, as although cost cutting which was essential to the firm's survival was aided by the way in which wages were reduced in the collective contracts signed by the Fascist unions, at Magneti, unlike many other firms in this early period, the economic problems were not simply faced by an all-out attack on the work-force's conditions and wages and at least the collective contract rates (low as they were) seem to have been respected. This is suggested by a report made by the Fascist unions in 1929 on how the various Milanese factories responded to the economic crisis, according to which, although some firms were ignoring the negotiated rates completely, at Magneti

Here the situation is better. The union representatives have room to breathe. It appears that the contract has been respected and that there has been a general pay rise for the women, who were previously mostly getting below the minimum rate.[14]

[13] 'Dalla Magneti Marelli', *Unità* (31 Mar. 1926), 3.

[14] N. Cuzzeri, (Ispettore sindacale), 'Promemoria. Situazione general operai metallurgici. Zona di Sesto San Giovanni', 12 Oct. 1929, ASM, Gab. Pref., cart. 1061, p. 2.

But this situation proved transitory and in the crisis of the early 1930s, when cost cutting became increasingly necessary, there was particular pressure on wages. One Communist Party report written in 1934 by a sympathizer who worked as a pay clerk at Magneti revealed some of the circumstances which rendered it difficult for workers to earn a decent wage. She described, for example, the fact that new women were paid according to a bonus system from the day they finished their eight-day trial period, criticizing this as unfair because the women were too inexperienced and often could not earn even the minimum percentage of 35 per cent. This clerk also relates how, when new products were introduced, often the times were not recalculated and frequently it was difficult to earn the bonus on the new work. Furthermore, various methods were found to keep the percentages down or even cut:

None of the women are allowed to earn more than 70 per cent. The Organizational Department checks the records of the percentages earned daily and the percentage rates on any operations where the workers have earned more than 70 per cent are immediately cut by half. Another thing that I noticed recently is the fact that an operation that was previously paid at a rate of L. 9,70 was subsequently cut to L. 2,45 and another from L. 6,30 to L. 3. A woman who could earn 100 per cent on her base rate of L. 0,82 per hour with the old rates, can't even earn enough to cover her daily expenses with the new ones. The bonus rates are calculated as if the hourly base pay were L. 0,82, yet sometimes, because of the needs of production, women who have a higher base rate are transferred there. If the lower paid only get 50 per cent with their low levels of pay, how can the others earn anything at all? In conclusion, every day there seem to be further cuts in the rates forcing the women to work even harder.[15]

RESISTING THE TIMEKEEPERS

Dissatisfied as many of them were with the manner in which the pay bonuses were calculated the majority of the workers were far too afraid to make any sort of public complaint:

[15] 'Rapporto di una simpatizzante alla Magneti Marelli', Jan. 1934, APCI, f. 1241, fo. 27.

There was no way to protest. You just had to take what was in your paypacket, put it away and say: 'It isn't much but I have to just take it, I can't protest.' (Pierino)

The report from the clerk quoted above reiterates this idea:

Amongst these workers paid according to the bonus incentive no one has the courage to protest because they are afraid of being discovered and threatened by the head of the shop.[16]

This fear was hardly surprising given the number of guards patrolling the shop-floor and the general level of discipline. Movement from shop to shop was easily controlled since workers in each shop had to wear different coloured overalls, and even fairly minor infractions were punished by fines or suspensions or even sacking.[17]

Nevertheless, in spite of such a repressive and highly disciplined atmosphere there is some evidence that anger over the pay system did lead to covert or sometimes even open resistance to the work of the timekeepers. In 1937, for example, *Unità* reported a stoppage over the question of pay: 'At Magneti Marelli, an entire shop protested against the fact that the bonus had not been paid, by stopping work.'[18] In 1939 another similar report appeared in *Lo Stato Operaio*. According to this,

On more than one occasion in the factory, the workers have agitated against wages cuts and for better treatment. On two of these occasions, in two different shops, the workers, to make their protest more effective, stopped the machinery.[19]

[16] 'Rapporto di una simpatizzante alla Magneti Marelli', Jan. 1934. APCI, f. 1241, fo. 27.

[17] e.g. in 1936 a man was suspended for 3 days without pay for smoking in the toilets and a similar fate met a woman who was discovered 'dancing with another woman during work hours'. Thefts, even of worthless materials, were severely punished. In 1944 a woman who had worked for the firm since 1929 was sacked on the spot for stealing some discarded pieces of metal to decorate a Christmas tree for her child. (These examples all come from reports contained in the personnel files in the APMM).

[18] 'Lotta per la pane', *Unità*, xiv. 7 (1937), 5. This stoppage was also mentioned in the testimony of one of the clandestine Communist Party members who helped organize it. See 'L'Antonio Mantovani', in Alasia, *La vita di prima*.

[19] 'Dati e considerazioni sulle condizioni della classe operaia italiana. Le condizioni di lavoro degli operai della "Magneti Marelli" ', *Lo Stato Operaio* (15 June 1939).

Some of these actions were provoked by workers trying to get 'justice' about the way in which their pay was calculated. Santino, who was both a foreman and a member of the clandestine Communist Party, remembers:

> We did actually have two or three demonstrations in the shop to protest against the fact that the times were too tight and the workers couldn't earn a bonus percentage because, for example, instead of having fifty seconds for a certain operation they only had forty. And so the workers couldn't earn their bonuses and that's why we had the protests. The workers—the women as well, there were a lot of women working on the smaller machines—went to the works manager. They went there, they protested and they stopped work. Thirty or forty of them went there to protest that they didn't agree with the times that the time study engineer had taken according to that Bedaux system. (Santino)

This evidence is all from the late 1930s and reports of actual strikes and agitations previous to this have proved impossible to find. The lack of evidence may not mean that similar stoppages did not occur, but for this earlier period it is only possible to trace a murmur of discontent through the fragmentary reports available, particularly in the period of the economic depression.

Maybe, in any case, to pay attention only to open protest is the wrong approach given the political situation. It may be more relevant to consider other more subtle methods that the workers could use in an attempt to thwart or at least retain some control over the bonus system. Of course, simply working harder was not enough to increase a pay packet. If productivity rose and everyone began to earn more then the bonus rates were simply cut. As in many other firms where bonus systems were introduced, numerous 'tricks' had to be used to ensure that the percentages were not cut and the work-pace raised.

One of these 'tricks' was the slowing down of the work-pace while the time study engineer was calculating the times and fixing the percentage rates. Maria G. said that

> a worker could try to show that a certain job took five minutes instead of two but she had to be very crafty to prove this. If she could do a five-minute job in two minutes then she could really earn well. They might, for example, expect her to get a 20 per cent bonus but in three minutes she could get 100 per cent. Of course she couldn't earn this all the time or they would just cut the rates.

Gina remembered that

> you really had to be quite crafty. For example, if I could finish something in ten minutes, if I worked slowly then I could take twenty minutes. That way we could keep our wages up by keeping the bonus rates up.

Another method used by the women in the Winding Shop was the hiding of finished work. If a worker produced a very high output she would hide some of her production so that it would not be apparent that such a large amount could be produced in a single day:

> we used to hide finished work. Then we could bring this out if we weren't feeling well, or if, without meaning to, we made a mess of the work we were doing and we had to stop a number of times to disentangle the wires. It was then that our secret store came in handy. (Carla)

MEN'S PAY, WOMEN'S PAY

The problems with the bonus system and the attempts to defy it were shared by both men and women alike. Yet the actual level of their wages was very different. This is immediately apparent from the examples in Appendix 4. In 1933, for example, Guidetta and Rosa were getting respectively only L. 10,64 and L. 9,52 per day. Certain other women earned considerably more, such as Ester who got L. 16,70 per day in the same year and Antonia who was getting L. 17,20 per day, but when compared with even the worst-paid men the women's wages were low. Even a very unskilled male worker like toilet cleaner Carlo was earning L. 24,00 in the same year, and a skilled male worker like the fitter Alberto was earning L. 29,32 which was three times Rosa's income.

It is evident that this was one of the main reasons why the firm employed so many women. Women, quite simply, suited the needs of capital because they were much cheaper than men. But it is difficult to tell how this disparity in male and female earnings was viewed by the work-force themselves. Equal pay for men and women was not on the political agenda in this period. Admittedly, it was occasionally championed by the leaders of the Fascist unions, but this was motivated by the fact that lower wages enabled women to undercut male workers rather than by any concern for the position of the women themselves.[20]

For many of the women themselves such low wages must have constituted a considerable practical problem, for, although sometimes a woman's wage might be seen as supplementary to her husband's or her father's, there were also many women like Olga (one of the women I interviewed) who were heads of families or who lived in families where they were the sole breadwinners. Her situation was not particularly unusual. Twenty-eight per cent of all families who migrated to Sesto San Giovanni between 1910 and 1920 were listed as having female heads in the Town Hall records.[21] Moreover, the question of wages cannot be judged simply by this yardstick alone, for given the sense of pride and respect that these women clearly ascribed to their ability to earn (see below), the fact that, however hard they worked, their wage packets were always much smaller than a man's, could only serve to reinforce their subordinate position in the family and workplace.

'ITALIAN AND FASCIST WAGES'

In general, wages at Magneti do not appear to have been any lower than in other engineering firms. Indeed, for the female workers they were fairly good compared with certain other possible jobs such in the textile industry. For the men, on the other hand, they were no higher than they could have obtained in other local engineering firms.[22] For most of the men, Magneti paid 'normal wages' for the time and the place, or as one ex-worker preferred to express it, Magneti paid its male workers 'Italian wages, Fascist wages' (Santino).

[20] Equal pay was proposed e.g. by Del Giudice who was Commissario della Confederazione dei Sindacati Fascisti del Commercio. See P. Meldini, *Sposa e madre esemplare* (Rimini–Florence, 1975).

[21] D. Bell, *Sesto San Giovanni: Workers, Culture and Politics in an Italian Town, 1880–1922* (New Brunswick, NJ, 1986), 242.

[22] One possible exception to this was the relatively good rate (for unskilled workers) accorded men in the B Plant, but their pay was only raised in order to 'compensate' for the dangerous, unhealthy working conditions there.

The fact which troubled the Magneti workers, therefore, seems to have been less the actual amounts they could earn (little as this may have been) than the fact that the bonus system was perceived as an elusive, incomprehensible factor over which they had no control and which was perpetually changing. It does seem that, despite Magneti's pioneering claim to be in the vanguard of productive, technological, and managerial techniques, particularly in the field of personnel management, when it came to the work-force's reaction to the pay system this firm was not unusual. The increased pace of work and the role of the growing Taylorist bureaucracy created to calculate standard times and bonus rates led to resentment. Here, as elsewhere, the timekeeper was seen as the worker's enemy to be outwitted if possible, and at times this general undercurrent of discontent erupted into various forms of resistance. In this respect at least, it appears that the firm's policy aiming to win the co-operation of the work-force for the introduction of scientific management was far from a total success. Ill feeling on this issue was seemingly widespread, although it mostly remained voiceless or muffled due to the severely disciplined atmosphere of the factory.

5

The Workers: Composition of the Work-Force and Labour Mobility 1919–1940

SINCE the publication of G. Procacci's classic work *La lotta di classe in Italia agli inizi del secolo XX* in 1972,[1] the composition of the industrial labour force has been a focus of interest among many historians of the Italian working class. One of the main problems with works such as Procacci's, however, has been the difficulty in finding adequate sources and the need to rely on official statistics such as censuses. More recently a different approach to the quantitative study of the composition of the Italian working class has been opened up with the statistical analysis of the payroll registers (*libri di matricola*) of specific factories, making use of computers to handle the large amounts of data involved. This work has, to a certain extent, been inspired by analogous research in other countries such as that of Rolande Trempé.[2] In Italy the first major study of this kind, published in 1983, was the highly interesting and extensive analysis of the inter-war labour market in the port of Marghera (near Venice). In addition there have been a number of studies of the work-forces of single plants or firms, such as the Meccanico di Sampierdarena plant of Ansaldo, the Savigliano steelworks, Alfa Romeo, the Cogne steelworks, Portoferraio, and the Ercole Marelli plant in Sesto San Giovanni.[3]

[1] G. Procacci, *La lotta di classe in Italia agli inizi del secolo XX* (Rome, 1972).

[2] R. Trempé, *Les Mineurs de Carmaux, 1848–1914*, 2 vols. (Paris, 1971), and 'Pour une meilleure connaissance de la classe ouvrière: l'utilisation des archives d'entreprises: le fichier du personnel', in *Mélanges d'histoire sociale offerts à Jean Maitron* (Paris, 1978). See also T. Hareven, *Family Time and Industrial Time: The Relationship between the Family and Work in a New England Community* (Cambridge, 1982).

[3] Belotti *et al.*, *I primi operai di Marghera: mercato, reclutamento, occupazione 1917–1940* (Venice, 1983); P. Rugafiori, *Uomini, macchine, capitali: l'Ansaldo durante il*

Although the *libri di matricola* of Magneti Marelli for the inter-war period have proved untraceable, I was able to make use of a far richer source of information.[4] In line with the firm's innovative attempts at 'scientific personnel management' a very detailed record was kept of each worker in the shape of individual files containing the original record card filled in for each new recruit.[5] Using a sample of these files[6] it was possible to investigate the composition of the Magneti labour force and draw at least some tentative conclusions about which kind of workers were employed in such a self-consciously modern factory. Were they, for example, mostly male or female, young or old, of rural or urban origin, skilled or unskilled? In particular, an important aim of this study was to compare and contrast the situation of male and female workers. As the following sections show, this differed greatly in terms of a whole series of factors such as age, skill composition, level of education, and place of residence. Furthermore, this analysis of the labour force of Magneti Marelli has proved a challenging

fascismo 1922–1945 (Milan, 1981); A. Donvito and G. Garbarini, 'Senz'altra formalità che il reciproco preavviso: le officine di Savigliano, 1904–1914', *Italia contemporanea*, 157 (1984); C. Binel, *Dall'Ansaldo alla Cogne: un'esempio di siderurgia integrale (1917–1945)* (Milan, 1985); D. Bigazzi, ' "I più turbolenti della città": la composizione operaia all'Alfa Romeo', R. Covino and G. Gallo, 'La forza lavoro della fabbrica d'armi di Terni durante la prima guerra mondiale', and P. Rugafiori, 'Occupazione e composizione operaia all'Ansaldo', all three articles in G. Procacci (ed.), *Stato e classe operaia durante la prima guerra mondiale* (Milan, 1983); C. Biffoli and M. Lungonelli, 'Una classe operaia in formazione: i siderurgici di Portoferraio (1901–1905)', *Studi storici*, 1 (1985), 53–68; S. Licini, 'Operai a Sesto San Giovanni: composizione e mobilità della forza lavoro alla Ercole Marelli (1929–1945)', *Storia in Lombardia*, 3 (1987); D. Bigazzi, *Il Portello* (Milan, 1988).

[4] I should like to thank the staff of the Personnel Department of Magneti Marelli (especially Dottoressa Bossi) for granting me access to these materials and for all their generous help and assistance during the months that I spent researching in the factory archive.

[5] Most of the files also contained a varying amount of other data, including, for many of the workers, regular reports about pay, work hours, promotions, ability, health, and disciplinary matters.

[6] The statistical analysis looked at a sample of those who started work at one of the Sesto Magneti plants or the N (in nearby Crescenzago) between the founding of the firm and the entry of Italy into the Second World War (10 June 1940). It does not include the work-forces of other firms in the Magneti group such as FIVRE, nor does it include clerks or managers. See Appendix 2 for more detailed information on the methodology employed in this statistical analysis. This appendix also contains numerous additional tables.

contribution to recent historiographical debates about labour mobility in the inter-war period and demonstrates the usefulness of focusing on gender, an aspect little considered in many other analogous studies.

THE MAGNETI WORK-FORCE: SEX AND AGE

Although Magneti Marelli differed from other Sesto engineering firms in its far larger numbers of female employees, these by no means made up the majority of the work-force, and looking at patterns of recruitment alone they seem startlingly few given the reputation of this firm locally as a major employer of women. During the inter-war period 35.4 per cent of new recruits were female and 64.6 per cent were male. However, these figures are misleading as the proportion of women in the actual work-force was considerably higher (nearer a half) given that, once recruited, women tended to stay longer in the firm. The sex composition of annual recruitment also varied over time (see Table A, Appendix 2), reflecting both the effects of wars and rearmament and the rise or decline of certain product lines. The expansion of some products such as radios led to the increased recruitment of women whereas others, such as accumulator batteries, were manufactured almost entirely by male labour.

Less variable over time was the average age of new recruits, which tended to be consistently fairly low throughout the whole inter-war period. Although some evidence suggests that this was the result of a deliberate management policy[7] (as the young often made stronger and quicker workers), it may have equally been due to a plentiful supply of young labour locally. The average age was also kept down by the number of women in the work-force, as female recruits tended to be somewhat younger (see Table 5.1). About a third were under 18 and nearly half

[7] e.g. an anguished letter written in 1935 to Soliveri, the works manager of the A Plant, from an unsuccessful female applicant for employment, who even had a recommendation from the local *fascio*, contained the following plea: 'Unfortunately, because of my age (38) you decided not to employ me. I beg you, sir, to see if it is possible to make an exception to the rule and give me some work in view of my desperate situation. For this reason I am willing to accept any kind of work you can offer as I cannot find employment anywhere.' (Letter to Michele Soliveri, 5 Oct. 1935, APMM).

Table 5.1. Age of new recruits

Age (years)	Women		Men	
	No.	%	No.	%
12–14	25	4.0	39	3.4
15–17	164	26.2	129	11.2
18–19	98	15.7	81	7.0
20–25	200	32.0	348	30.2
26–35	114	18.2	351	30.5
Over 35	24	3.8	204	17.7
	625		1,152	

Note: figures in all tables refer to the sample and not the entire work-force. Figures for the male sample are weighted by a factor of three (see Appendix 2).

were less than 20 years of age. In comparison only 14.6 per cent of the male workers were under 18 and only 21.6 per cent under 20. In addition, almost half of the men were over 25 compared with 22 per cent of the new female recruits. Nevertheless, as the latter statistic reveals, although the women tended to be younger than the men, they were far from all young girls.

SKILL

The question of evaluating the skill composition of a work-force has often proved difficult. For example, the Porto Marghera, Alfa Romeo, and Savigliano studies had to rely on job names to assess the grade of each worker, a clearly problematic method without a very detailed understanding of what each job entailed. In contrast the Magneti archive contains the pay grade in addition to job names and it seemed more useful to use the former as a measure of the level of skill. In other words I have used the skill categories of the collective contracts negotiated by the Fascist unions.[8]

[8] Unfortunately, information is lacking in many of the files, especially those for the 1920s, although for the 1930s there is enough data to make possible a tentative analysis of the skill composition of the Magneti recruits.

The whole concept of skill is, of course, historically and culturally defined. The reasons for calling one job skilled and another unskilled often have more to do with the political muscle of certain sections of the labour force or with general cultural values than with some intrinsic aspect of the work itself. This is particularly true of women's work, since once a woman is put onto a certain job it is frequently assumed that this implies that the job is unskilled or semi-skilled.[9] As might be expected in such an 'Americanized' firm, clearly much of the work at Magneti did seem (as far as this can be understood from the pay grades) to have had a low skill content in comparison with other local firms.[10]

In the inter-war period about half of the Magneti men and many of the women were doing what was classed as unskilled or semi-skilled work. There was a core of skilled male workers, doing highly specialized tasks, and a much larger number of unskilled and semi-skilled workers of both sexes doing a variety of jobs. The women and the younger workers were largely confined to the A Plant (and later the N), where they composed the majority of the work-force. In the B, on the other hand, where work was heavy and dirty, most of the work-force were unskilled, adult men (see Table B, Appendix 2).[11]

[9] On definitions of skill and women's work see e.g. A. Philips and B. Taylor, 'Sex and Skill: Notes towards a Feminist Economics', *Feminist Review*, 6 (1980).

[10] This point is less obvious than it may appear, given that the classic work of Harry Braverman which equated scientific management with a relentless process of deskilling has since been challenged. Empirical studies have shown that technological and organizational change leads to a recomposition of skills rather than to a simple process of downgrading. See H. Braverman, *Labor and Monopoly Capital: The Degradation of Work in the Twentieth Century* (New York, 1974). For a critique of Braverman's deskilling thesis see, for example, B. Jones, 'Destruction or Redistribution of Engineering Skills? The Case of Numerical Control', in S. Wood (ed.), *The Degradation of Work? Skill, Deskilling and the Labour Process* (London, 1983).

[11] This table must be treated with care, as only about a third of the sample had both grade and plant noted on their files. The information is, therefore, partial. However, the fact that the work-force of the B Plant differed in this way is backed up both by oral testimony and by statistics for 1941 collected by the governmental 'osservatori industriali' (industrial spies). On 31 Dec. 1941 the work-force of the B Plant was reported as follows: 1 manager, 18 male clerical workers, 12 female clerical workers, 10 *operai specializzati*, 56 *operai qualificati*, 168 *manovali specializzati*, 26 apprentices, 24 women, and 4 boys (total work-force 320). (ACS, OI, b. 22, f. 1, 'FIMM Stabilimento B'.)

Female recruits were largely undifferentiated by grade in this period (see Table C, Appendix 2) and many were simply classified as either *ragazze* (girls) because of their age, or as apprentices who might be young or old. By the later 1930s the practice of initially calling women apprentices seems to have declined and a growing proportion were simply graded as B workers from the moment they started work in the factory. Only a tiny proportion started in the more highly paid A grade.

In comparison the male workers were divided into a greater number of grades, reflecting the greater variety of roles they had within the factory. About a quarter of the men were recruited as skilled workers (*operai specializzati* and *operai qualificati*), a further quarter as boys or apprentices, and the remaining half were classed as unskilled or semi-skilled (*manovali*). The figures for the 1920s are, as for the women, so few as to be of dubious significance, but during the 1930s the main trend seems to have been a considerable rise in the proportion of semi-skilled workers (*manovali specializzati*), their numbers rising from 22.5 per cent of the male recruits in the period 1931–5 to 44.9 per cent during 1936–40. This may be clear evidence of 'deskilling' as a result of the spread of rationalized production methods requiring increasing numbers of semi-skilled workers who could be used, for example, on automatic machines. Alternatively (or additionally) this phenomenon may be seen as part of a more general trend in Italian industry, amply documented in the antifascist press, to reclassify jobs downwards in order to cut wage bills, a trend facilitated greatly by the absence of real trade unions. At Magneti, the move to downgrade jobs affected existing workers as well as the pattern of new recruitment, and some of the workers' files recorded a drop in grade.

LEVEL OF EDUCATION

As with their age and skill classifications, in terms of their educational background the situation of the male and female workers proved to contrast greatly. In general, this was a fairly well educated work-force. Illiteracy was impressively low, affecting only 0.7 per cent of the women and 1 per cent of the men.[12] Some of the workers even had a certain level of technical education such as the new *scuola di avviamento al lavoro*,[13] although most had not gone beyond primary level. But as Table 5.2 reveals, there were many differences between the level of education of male and female workers. The overwhelming majority of women (97.3 per cent) had only primary education compared with 59.2 per cent of the men. Furthermore, of those who had had only primary education, the men were more likely to have completed a greater number of years. During the whole inter-war period the educational level of the female recruits barely altered, whereas the level rose steadily for the men (see Table D, Appendix 2). This probably reflected less a deliberate strategy to recruit better-educated workers than more generally the rising standard of boys' education.

On the whole, women's educational level was low regardless of place of birth or residence. Men's, on the other hand, varied

Table 5.2. Level of education of recruits

Level of education	Women		Men		Total
	No.	%	No.	%	
Illiterate	2	0.7	6	1.0	8
2 years primary	9	3.3	6	1.0	15
3 years primary	51	18.9	51	8.3	102
4 years primary	75	27.8	78	12.7	153
5 years primary	54	20.0	99	16.2	153
6 years primary	21	7.8	51	8.3	72
Primary (? no. of years)	52	19.3	78	12.7	130
Secondary	6	2.2	243	39.8	249
	270		612		882

[12] In the province of Milan in 1931 the overall illiteracy rate for the population over 6 years of age was 4.2% (3.0% in the city of Milan and 5.6% in the rest of the province). This constituted a considerable improvement on the 1921 figures (7% for the city and 9.3% for the rest of the province). See G. Della Valentina, 'Agricoltura e aspetti del rapporto città campagna in Lombardia tra crisi e seconda guerra mondiale', in *Società e storia*, 20 (1983), 342.

[13] On the various changes in legislation and working-class education see F. Cereja, 'La scuola e il mondo del lavoro', in *Annali Feltrinelli* (1979–80), 51–90.

more. Rural men were less well educated than their urban counterparts, and unsurprisingly those who started work in a skilled grade were better educated than the *manovali* (see Table E, Appendix 2). This underlines one fundamental point about the differing situations of men and women in the factory. For men there was a chance of promotion and some did skilled work where a technical training might be of use, whereas women's work was generally either unskilled or semi-skilled. If a woman did manage to stay on at school above the primary level she was unlikely to seek this type of factory work, since there might be a possibility of a clerical position.

However, this is not the only explanation for the lower educational level of women. Boys in general were far more likely to be allowed by their families to stay on at school or encouraged to attend evening classes. In spite of the fact that women generally did waged labour of some sort throughout their adult lives, only boys were expected to aspire to a 'career'. For example, Maria G., one of the ex-Magneti workers I interviewed, described how her father went to Germany during the war: 'he used to send home 1,000 lire a month or even more . . . He went there so my brother would be able to stay on at school.' This desire to have the best for their son contrasts sharply with her parents' attitude to sending their daughter to work. She was sent immediately at the legal minimum age, 'because if they could have sent me to work even earlier than that, they would have done so'.

Gender was, therefore, an important divide in the educational level of the work-force, since these women's education was limited regardless of factors such as geographical origin, job in the factory, or age, whereas men's education reflected the more varied composition of the male work-force and the greater opportunities open to them.

MARITAL AND PARENTAL STATUS

The majority of the Magneti recruits were unmarried, and many of the married ones were childless (see Table 5.3). This was especially true for the women, 74.4 per cent of whom were unmarried, as opposed to 61.7 per cent of the men. Furthermore,

Table 5.3. Marital and parental status of recruits

Status	Women		Men		Total	
	No.	%	No.	%	No.	%
Unmarried	308	74.4	618	61.7	926	65.4
Widowed/separated	8	1.9	6	0.6	14	0.9
Married, childless	52	12.6	102	10.1	154	10.9
Married, 1 child	32	7.7	117	11.7	149	10.5
Married, 2 children	12	2.9	96	9.6	108	7.6
Married, 3–7 children	2	0.5	63	6.3	65	4.7
	414		1,002		1,416	

of the married men, 73 per cent already had children, compared with 46.9 per cent of the married women[14] (see Table F, Appendix 2). Unfortunately marital status is noted only on the *foglio d'assunzione* and there is no indication as to whether those who started work unmarried subsequently married, or whether they had children. For women, marriage or motherhood did not necessarily mean leaving the labour market, but it could mean a change of occupation, since factory work might be difficult with small children to care for. At almost all points in their lives women worked to earn money in some capacity, but once they became mothers this might be sewing or laundry which could be done at home. Few working-class families could manage on one income alone, especially with children to feed.[15]

The question here, therefore, is not whether women stopped working but whether they had to give up factory work. In the majority of cases, marriage itself seems to have made no

[14] These figures only reflect what was noted in the actual files. Of course, in reality, some of the unmarried workers may have had children too, but preferred not to declare them.

[15] e.g. the testimonies of working-class women collected by Bianca Guidetti Serra in her book *Compagne* (Turin, 1977) well illustrate the fact that women worked all their lives in this period. See also C. Saraceno, 'La famiglia operaia sotto il fascismo', in *Annali Feltrinelli* (1979–80), 210.

difference and women remained in the firm as before. Maternity was, on the other hand, sometimes given as a reason for leaving the firm in some of the women's files, although it is hard to tell whether such a cause was frequent due to the scarcity of the data (see Table G, Appendix 2). It is not possible to know directly, therefore, whether women usually left the factory on the birth of a first child. One indirect approach is to look at the age of women when they left Magneti. Given that the average age of marriage for women was about 24 in this period,[16] a good number of the women may have already had at least one child by the time they left the firm. The average age on leaving the factory was not especially low; between 1927 and 1940 the mean age of women leaving was 27.7 years compared with 32.1 years for men (see Table H, Appendix 2).

A large family would have been a very difficult burden for female factory workers, but with only one or even two children it was possible for women to continue work at Magneti. The firm provided a nursery and nursing facilities, and since the majority of the women lived close to the workplace many of them were probably able to make use of these.[17] Other women were helped by relatives, frequently the child's grandmother, to care for the child during work hours. Carla, for example, one of the women I interviewed, relied on her mother to care for the baby while she was at work and in the early months after birth the grandmother came to the factory at regular intervals so that Carla could feed her child in the breast-feeding room.

PLACE OF BIRTH

As Anna Treves has argued in her study of internal migration in

[16] Mean age at first marriage of females:

	1923–5	1926–8	1929–31
Lombardy	24.3	24.2	24.2
Italy	24.0	23.9	23.9

Source: M. Livi Bacci, *A History of Italian Fertility During the Last Two Centuries* (Princeton, NJ, 1977), 107.

[17] 80% of the women in the sample who already had children when they started work lived in Sesto itself or in the area of Milan next to the factory. This figure is not, however, substantially different from that for the unmarried women, 72.6% of whom lived in the same area.

the Fascist period,[18] in spite of the Fascist ruralization campaign and legislation, the inter-war years saw considerable movements of people within Italy. Emigration to America had slowed down to a trickle due to the 'Quota Act' of 1921, and during this period emigration (mostly of northern Italians) focused on other European nations. Internal migration occurred not only between villages and towns in the same area, but over much longer distances, often from poor areas such as Veneto to the large industrial cities.

In this period Milan and Sesto both experienced considerable immigration, mainly from Veneto, Friuli, and the south. Migrants had already been flooding into Sesto from the early twentieth century as it developed into an important manufacturing centre[19] and this growth continued in the Fascist period, with the resident population rising from 18,274 at the 1921 census to 35,845 by August 1934 (see Table 1, Appendix 3). Initially, the new arrivals tended to work in the heaviest jobs such as labouring on the regime's public works, such as Milan's new Central Station, or in factories with many health risks like the rayon-making firm Snia Viscosa. Many women became domestic servants.[20] But some of these migrants found work in engineering firms like Magneti, and almost every region of Italy was mentioned on the workers' files as their place of birth. Only about half (47.5 per cent) were actually born in the province of Milan, and almost a third (30.6 per cent) came from outside Lombardy. Maybe, as in the nearby Breda factory, the management deliberately recruited immigrant labour,[21] but no evidence of this has emerged. It is important to remember, furthermore, that the files give only the place of birth, but no indication of when or how a worker moved to Milan. There is a distinct difference between adults who moved in search of work and

[18] A. Treves, *Le migrazioni interne nell'Italia Fascista* (Turin, 1976).

[19] On this changeover see L. Bosisio, 'Gli insediamenti industriali e l'urbanistica della transizione, 1880–1914', in G. Petrillo (ed.), *La città delle fabbriche: Sesto San Giovanni 1880–1945* (Milan, 1981). As Bosisio notes, this rapid expansion caused innumerable problems as housing and other services lagged behind the rate of growth; for example, Sesto only got a sewage system in 1930.

[20] G. Consonni and G. Tonon, 'Milano, classe e metropoli tra due economie di guerra', *Annali Feltrinelli* (1979–80), 440.

[21] 'In 1926, at Breda in Sesto San Giovanni, the management was sacking and laying off workers on the one hand whilst simultaneously recruiting new ones from Friuli or the South', ibid. 439.

those who saw themselves only partly as immigrants, having moved with their families as children to the Sesto/Milan area.

In 1926 an article in *Unità*, the communist newspaper, depicted the typical Magneti worker as a desperate, impoverished immigrant from the poorer parts of Italy:

> the majority of the workers are immigrants from the South or from the Veneto area. They usually are not very good workers (hardly surprising considering how hungry they are!) and will accept any conditions and the lowest rates of pay in order to survive.[22]

It might appear self-evident that in such a factory as this newly-arrived migrants would serve as a useful pool of cheap, unskilled, and semi-skilled labour. It is true that in the 1920s 22.7 per cent of the workers did come from Emilia and Venetia and a further 6.3 per cent came from the centre or the south. However, the writer was most probably referring to the situation in one part of the factory alone, for it would be incorrect to state that migrants constituted the majority of the work-force. In addition, when the relationship between workers' grades and their place of birth is examined, no significant link is apparent. This lack of a clear pattern is in itself worthy of consideration. It reveals that the reality was somewhat more complex than the *Unità* correspondent suggested. No consistent figure of a typical migrant emerges. Some probably were indeed the unskilled, rural poor moving to a large city to escape a terrible poverty cycle. But just as often they might be skilled workers with at least some capacity to choose where they wanted to work. They had a skill to sell and were more likely to have savings to finance the move. Furthermore, it is possible that many of these were, in practice, 'second generation immigrants' even though they were actually born in other regions of Italy.

A number of the ex-workers I interviewed were themselves born in other regions. Crucial to the move to Milan was, in many cases, the presence of cousins, aunts, uncles, or other relatives already resident in the city who could find accommodation for the migrants or provide initial hospitality. The classic

[22] Il corrispondente d'officina, 'Dalla Magneti Marelli di Sesto San Giovanni', *Unità* (31 Mar. 1926), 3.

pattern of chain migration is evident: frequently, one family member was sent first to stay with a relative and seek work; later the rest of the family would join her or him. One ex-worker, for example, recalled how she came to Milan initially to learn dressmaking and then returned to her village. Later she went back to Milan with her family, relying on the extended family network:

My whole family came here, because my uncle who lived in Milan (and I had another one too who was here in Sesto itself, working as a clerk in the Town Hall) told us we could all come. First of all my uncle from Milan found us somewhere to live. Almost all my father's brothers lived here in the Milan area. (Germana)

Where then did the Magneti workers come from? As Table 5.4 shows, only a small percentage were actually Sesto born, which is hardly surprising in view of the expansion of Sesto's population due to immigration in this period. Over two-thirds of the workers (69.3 per cent) came from Lombardy including Milan and Sesto, many of whom were probably migrants who

Table 5.4. Place of birth

Place of birth	Women		Men		Total	
	No.	%	No.	%	No.	%
Sesto San Giovanni	75	14.5	63	5.7	138	8.5
Milan	34	6.6	123	11.1	157	9.7
Within 15 km	127	24.5	219	19.8	346	21.3
Over 15 km, Milan prov.	40	7.7	90	8.2	130	8.0
Bergamo, Brescia, Como	63	12.2	147	13.3	210	12.9
Rest of Lombardy	36	7.0	108	9.7	144	8.9
Emilia	32	6.2	69	6.3	101	6.2
Venetia	58	11.2	111	10.1	169	10.4
Piedmont and Liguria	11	2.1	36	3.3	47	2.9
Centre and South	39	7.5	108	9.8	147	9.1
Abroad	3	0.6	30	2.7	33	2.0
Total	518		1,104		1,622	

had moved, at some point, from the poorer rural areas to the more industrialized zones. Apart from the 12.9 per cent who had come down from the largely mountainous provinces of Bergamo, Brescia, and Como to the north of Milan, the main sources of migration were Emilia and the poorer peasant areas of Venetia and, to a lesser extent, the centre and south. It seems safe, therefore, to conclude that a good proportion of the Magneti workers were rural born, although it is not possible to tell what percentage of these were actually brought up in their place of birth.

This pattern seems to have altered little over time (see Table I, Appendix 2). Some trends are discernible, namely a decline in the proportion of Sesto and Milan born workers in the 1930s, an increase in those originating from the mountainous provinces to the north of Milan, and a rise in the numbers from the centre and south—but none are particularly marked. Neither do there seem to have been any significant differences between the origin of male and female workers; the pattern is remarkably similar. This seems only to reinforce my previous argument that migration was by families rather than individuals, so that women were as likely as men to be immigrants.[23]

Overall, there do not seem to have been significant differences as regards sex or grade between the immigrant and local-born workers. Neither does the pattern of immigration seem to have varied to any great extent over time. These conclusions are perhaps surprising in view of the widely held view that immigrants took the most menial, ill-paid jobs. At Magneti the workers' place of birth does not appear to have been a particularly significant factor. Without a better understanding of the point in their lives at which they migrated, however, it is not possible to tell whether the majority of those not born locally were, in fact, recent migrants or if they had spent much of their lives in the Milan area. The geographical origin of the workers in

[23] This pattern of whole families moving together existed from the early period of Sesto's expansion. One demographic study suggests that in the period 1900–20, although slightly more immigrants were male than female, the typical immigrants into Sesto were not young, single men but youngish married couples often with small children. See A. Morra and P. Sala, 'In-migration, Population Turnover, Industrialization: Sesto San Giovanni, 1900–1920', *Contribution of Italian Scholars: International Union for the Scientific Study of Population. XX General Conference* (Rome, 1985).

terms of their place of birth can give, therefore, only a limited understanding of the structure of the labour market. Far more revealing is an analysis of where the workers actually lived. Were they rural or urban resident, and how far from the factory did they live?

PLACE OF RESIDENCE

Although Sesto San Giovanni and the immediately surrounding towns of the Milanese hinterland such as Cologno Monzese grew dramatically in this period they could not provide enough labour for the expanding industrial sector. Furthermore, conditions in the countryside and the crisis in the textile industry encouraged peasants to seek employment further afield, attracted by higher urban wages. During the 1930s approximately one-fifth of the entire work-force of Milan was composed of commuters from poorer rural areas.[24]

Although much of this commuting was stimulated by the demand for cheap labour in the building industry, some of these rural workers managed to get employment in factories like Magneti. There is evidence that, in 1925–6 at least, this firm was deliberately recruiting in the poorer rural areas. According to the management this was necessitated by the labour shortage in this period: 'We recruit our workers (who are now almost impossible to find in the big cities) in the countryside and we bring them to the firm by a good, comfortable bus service.'[25] But the antifascist press offers a different explanation for this busing in of workers. Rural workers could be cheaper to employ:

> Since last August about a hundred girls have been bused in from the villages of Canonica d'Adda, Cassano, and Fara d'Adda. These girls are collected in the morning and taken back to their villages in the evening. Although they do exactly the same work as the others, they are paid a

[24] Consonni and Tonon, 'Milano, classe e metropoli', 444. See also Della Valentina, 'Agricoltura e aspetti', 338. Della Valentina argues that this increase in commuting had the effect of diminishing the differences between rural and urban areas, as is evidenced, for example, by the decline of rural 'superstitions' when the traditionally peasant world came into contact with the industrial city.

[25] *SB* 8 (1925), 2.

daily fixed rate of 9 to 10,50 lire. Now their numbers have been reduced from a hundred to only about forty and even these remaining jobs are threatened. These women have been enticed away from the mills by the prospect of a slightly higher rate of pay, but now, as the firm doesn't need them any more, they are going to be sacked and it will be impossible for them to return to their former jobs. Only unemployment awaits them.[26]

In spite of management strategies of this kind, as Table 5.5 shows, the majority of the workers employed at Magneti were resident in Sesto, in Milan, or very close to Sesto. The pattern of residence varied markedly according to sex. A very high proportion of the women (54 per cent) lived in Sesto compared with only 23.2 per cent of the men. Conversely, the male workers were much more likely to come from the city of Milan (27.1 per cent of the men compared with 4.1 per cent of the women). A slightly greater percentage of the men were rural

Table 5.5. Place of residence

Place of residence	Women		Men		Total
	No.	%	No.	%	
Sesto San Giovanni	291	54.0	258	23.2	549
Adjacent towns*	115	21.3	186	16.8	301
Monza	59	10.9	141	12.7	200
5–10 km from factory	9	1.7	30	2.7	39
10–15 km from factory	9	1.7	21	1.9	30
Over 15 km from factory	34	6.3	171	15.5	205
City of Milan	22	4.1	300	27.1	322
	539		1,107		1,646

* Cologno Monzese, Cinisello, Bresso, Vimodrome, Brugherio, Segrate, and the zone of Milan next to the factory.

[26] Il corrispondente d'officina, 'Dalla Magneti Marelli di Sesto San Giovanni', 3.

resident, and 15.5 per cent of the men compared with 6.3 per cent of the women travelled over fifteen kilometres to work. This picture remained substantially the same throughout this period for the men, whereas among the women there was some change as the percentage of locally resident women declined slightly over time and the number travelling from further afield rose (see Table J, Appendix 2). The explanation for the contrasting patterns of residence for men and women reflects both supply and demand on the local labour market. It also reflects the differing relationship between home and work for the two sexes.

Sesto was essentially a heavy engineering town with firms like Breda which employed large numbers of male workers. Locally the opportunities for women were far more limited, although a handful of factories did employ female labour, such as the Pirelli rubber works, the packaging firm Scatolificio Ambrosiano, and a small ribbon factory. As jobs for men were on the whole more plentiful, it was easier for the firm to recruit women locally. Even if the notorious inaccuracies of census data (with their tendency to underestimate women's role in the economy) are taken into consideration, Table 2 of Appendix 3 still constitutes a striking testimony to the difference in the labour market for men and women at the beginning of the inter-war period. Men overwhelmingly dominated the main fields of employment. In the largest local industrial sectors, which were engineering/metalworking and mineral processing/building, in 1921 they represented 94.93 per cent and 86.83 per cent of the work-force respectively. Women predominated only in the far smaller chemical and textile industries in the town.

The local labour market for men meant that it was not always possible for the firm to recruit the type of male labour it needed in Sesto. Skilled labour was often in short supply and, to a certain extent, skilled male workers could choose where they wanted to work. Unskilled male labour was more abundant, especially in the years of the economic crisis when unemployment was an important problem even in very industrialized towns like Sesto. Poverty in the countryside drove many to seek work in the great industrialized plants of Sesto, and queues of hopefuls outside factory gates were a common sight (and a source of concern to the authorities, since they made unemploy-

Table 5.6. Initial grade according to place of residence of male workers

Initial grade	Sesto		Milan		'Rural'		Total
	No.	%	No.	%	No.	%	
Op. specializzato	12	16.0	39	52.0	24	32.0	75
Op. qualificato	12	16.7	15	20.8	45	62.5	72
Apprentices	12	20.0	18	30.0	30	50.0	60
Manovali	69	23.5	60	20.4	165	56.1	294
Boys	33	39.3	9	10.7	42	50.0	84
							585

ment a visible problem and a threat to public order).[27] Table 5.6 shows that the worker's grades and geographical origin were, indeed, to a certain extent, related. The most notable feature of this table is the fact that more than half the most highly skilled male workers (*operai specializzati*) lived in Milan, whereas many of the unskilled lived close to the factory in Sesto or in nearby rural areas. This table must, however, be treated with caution, since only about half of the files had the grades noted.

This picture of a contrasting labour market according to sex

[27] The local authorities seemed more worried about the visibility of unemployment than about the problem of the lack of work itself. Groups of unemployed people waiting at the factory gates could lead to dangerous discussions: 'It is a general belief that currently there is an extremely high rate of unemployment in the city. This is because certain kinds of unemployed workers have got into the habit of waiting in long queues outside the gates of the largest firms in the city such as 'Pirelli', 'Stigler', and 'Miani e Silvestri', etc. In spite of the protests of the managers of the factories concerned who tell them there is no work, the unemployed do not go away and hang around for hours huddled together in groups, debating amongst themselves, and frequently criticizing the Party, the Fascist unions, etc. This makes the problem of unemployment publicly visible'. (Unsigned letter from the Milan Prefecture to the Minister of the Interior, 10 June 1930, ACS, PS. 1930–1, b. 54, f. 'Disoccupazione 1930'.) Since the problem was seen in terms of public order rather than of jobs, the solution was to make the unemployed less visible: 'The situation in the industries in the Sesto San Giovanni area, as in many other areas, is worsening and there will soon be numerous redundancies. Since letting these workers hang around waiting for jobs that might not turn up constitutes a threat to public safety it will be necessary to deport many of them back to the rural areas they came from . . .' (Prefect of Milan cited in a letter from the Direttore Capo della Divisione Politica to the Direzione Generale della Pubblica Sicurezza, 3 May 1927, ACS, PS, 1927, f. 'Milano', sf. 'Agitazioni operai'.)

and level of skill is further reinforced by numerous letters, conserved in the factory archive, from workers applying for jobs at Magneti. In the letters from male skilled workers there were often references to advertisements in newspapers such as the *Corriere della Sera*. For unskilled men and for women, on the other hand, the firm does not seem to have had the need to advertise and was able to rely on either workers sent by the Fascist Labour Exchange or the placing of boards outside the factory listing which type of workers were to be recruited that day.

The local supply of different types of labour is not, however, enough to explain adequately the much greater tendency for women to be resident closer to the factory, since unskilled male and female labour could both be recruited fairly easily in Sesto. There were also social and cultural reasons why most of the women lived near to the firm. Women were reluctant to travel long distances to work. This was partly due to parental disapproval of girls going too far from the home because of the dangers of travelling alone, and also to the dual burden of factory work and housework and sometimes even child care. Only one of the women I interviewed, Maria V., travelled a good distance to work (about twelve kilometres) and, as she herself remembers, it was exceptional in her village and not considered entirely respectable for a woman to be going so far away every day:

> I remember when I got married, people thought I was practically a prostitute just because I worked in Sesto: 'Oh, she works in Sesto. Who knows what kind of a girl she is!' Because there was quite a lot of local industry and hardly any of the girls worked outside the village. Nowadays, lots of girls work outside but in those days you could count them on one hand. Everyone was a bit shocked and they used to say: 'Oh my God! She goes all the way to Sesto!' So you see in those days I didn't really have many friends, most of the workers on the tram were men and there were very few women.

The geographical pattern of residence of the work-force demonstrates how different was the situation of men and women in the labour market. Women had little opportunity of factory work locally, other than at Magneti, and rural women were unwilling or unable to commute into Sesto, so that most of the female Magneti workers lived very near to their place of work. For men the situation was more complex, with the local and rural residents being a mix of different types of workers and many of the older, skilled workers coming from the city of Milan. This contrasting labour market for the two sexes was also important in determining the pattern of labour mobility.

LABOUR MOBILITY

Prior to the publication of the study of the inter-war labour market in Porto Marghera, *I primi operai di Marghera,* many historians believed that a high turnover of industrial labour was only a feature of the very early phase of industrialization and that once large-scale 'modern' industry emerged it led to the creation of a 'modern' industrial work-force. However, in Marghera's large factories of Breda, Montecatini, Ilva, and Sirma[28] the labour force was discovered to be very unstable and this and other analogous research suggests that a high turnover may have existed in Italian industry for most of the first half of the twentieth century. The Marghera research group argue that this was not simply a continuation of the pattern of nineteenth-century factories, such as textile mills, where industry was subordinated to the agricultural cycle and workers moved to and from agriculture and industry according to the season:

> in the case we have studied (and this seems to be true in other areas and not just Marghera) the instability is due, to a great extent, to the labour supply. Throughout the entire twenty years under consideration this high rate of mobility was caused and maintained by the availability of a huge pool of labour.[29]

This study concludes that the rate of turnover varied according to the type of worker—skilled or unskilled, rural or urban, and so on. The rural-resident unskilled worker, for example, was identified as particularly unstable on the industrial labour market:

> in rural areas this instability could mean that workers left industrial work either temporarily or definitively: in some cases, instability meant

[28] The type of production in these plants was: shipbuilding (Breda), chemical fertilizers (Montecatini), steelworks (Ilva), and non-inflammable materials (Sirma).

[29] Belotti *et al.*, *I primi operai di Marghera*, 170.

that when a particular job ended the workers still continued to look specifically for factory work which they considered to be their main source of income: in other cases, however, the factory was simply one of many ways of earning money but was neither the most favoured nor the most common.[30]

However,

> To a certain extent, this high rate of mobility affected both skilled and unskilled workers, and this suggests that we need to reconsider how to evaluate the relative positions of different 'worker figures'.[31]

From their investigation the Marghera researchers built up a list of the main types of workers in these factories, such as highly skilled craftsmen who remained in the same job for a long time, or urban workers who worked in one firm for only a short period, whether skilled or unskilled, or the different types of rural workers.[32] Less attention was paid to the significance of the sex of the workers under consideration, partly due to the fact that the firms studied employed few women.

Research on the work-forces of other Italian firms suggests that a high rate of labour mobility was widespread elsewhere. For example, Paride Rugafiori's study of the workers of one plant of the Genoa firm of Ansaldo, Duccio Bigazzi's study of the Alfa Romeo labour force, and Anna Donvito and Giovanni Garbarini's analysis of labour mobility at the Savigliano plant in Turin have all revealed a high turnover. The Marghera group may, therefore, be correct in arguing that

> the existence of large numbers of casual workers, working irregularly and forming an unstable proletariat, was not only characteristic of almost all of the Marghera industries but also an important feature of the rest of the industrial structure between 1920 and 1940.[33]

If this is the norm then Magneti Marelli should be considered the exception. It contrasted in many respects with the Marghera industries and firms such as Alfa Romeo. In particular it employed a far greater percentage of women and its management methods were more self-consciously 'modern'. The rate of turnover also differed greatly. At Magneti, labour mobility was comparatively low. To understand the reasons for this it is

[30] Ibid. 293.
[31] Ibid. 296.
[32] Ibid. 299.
[33] Ibid. 294.

necessary to consider a number of factors including management strategy, the composition of the labour force (especially as regards sex), the particular conditions of work in this firm, and the context of Sesto itself. Due to the rather exceptional nature of Magneti Marelli in comparison with the rest of Italian industry in this period my findings do not, of course, in any way negate the hypotheses of the Marghera group but rather provide an example of a contrasting situation.

As Table 5.7 shows, at Magneti Marelli in the inter-war period, less than a quarter of the work-force left before the end of their first year of employment and over 60 per cent stayed for more than two years. Table 5.8, which compares this with the firms studied by the Marghera group, highlights what a comparatively low turnover this constituted. It is clear, therefore, that for a good proportion of the Magneti work-force their experience of working in the firm lasted for a reasonable amount of time and they are not the largely transitory figures of the Marghera study, but industrial workers who identified themselves as such, stable in the same firm for a number of years. Why then was the rate of labour mobility relatively low in this firm? One factor which must be taken into account in explaining this phenomenon is the nature of the factory itself and in particular the work process and management strategy.

To a certain extent, the organization of the work process itself may have served to reduce labour mobility. According to an article in *Unità* in 1926:

> Because this firm uses mass-production methods, it means that the workers each specialize in doing only one very narrowly defined job. Often they can do nothing else. There have been, in fact, cases where workers have left to seek work elsewhere but have been forced to return and submit to even greater exploitation by this firm.[34]

How frequently this occurred is uncertain, but in any case the management did not trust to such factors alone and intervened more directly to retain its work-force. This was particularly important in the case of skilled workers who were hard to recruit, but management strategies such as the attention paid to the creation of welfare and leisure facilities suggest that in

[34] Il corrispondente d'officina, 'Dalla Magneti Marelli di Sesto San Giovanni', 3.

Table 5.7. Length of employment

Length of employment	Women		Men		Total	
	No.	%	No.	%	No.	%
Up to 6 months	52	9.4	159	16.5	211	13.9
6–12 months	48	8.6	114	11.8	162	10.7
1–2 years	65	11.7	147	15.3	212	14.0
2–5 years	156	28.1	249	25.9	405	26.7
5–10 years	150	27.0	138	14.3	288	19.0
10–20 years	62	11.2	120	12.5	182	12.0
20–39 years	22	4.0	36	3.7	58	3.8
	555		963		1,518	

Table 5.8. Length of employment in four firms

Length of employment	Percentage of work-force			
	Montecatini 1924–45	Ilva 1933–45	Breda 1928–45	Magneti 1919–40
Up to 1 month	35.0	6.5	7.1	4.4
1 month to 1 year	48.1	31.2	39.0	20.2
1–5 years	9.9	28.1	31.7	40.7
Over 5 years	7.0	34.2	22.2	34.8

Source: Breda, Ilva, and Montecatini information adapted from Belotti *et al.*, *I primi operai di Marghera: mercato, reclutamento, occupazione 1917–1940* (Venice, 1983), 288. The figures for Breda, therefore, refer to its Marghera rather than its Sesto plant.

general a low turnover was seen as desirable for the work-force as a whole. Furthermore, as health and safety hazards were not great in most of Magneti's work, the temptation to dismiss workers in order to replace them with new undamaged recruits was less relevant in this firm.

But management strategy alone is not sufficient to explain the pattern of labour mobility in this firm. In addition it is necessary to look at the attitude of the workers themselves and attempt to assess which factors influenced their decisions to remain or stay in the firm. It might be expected that some sort of insight into this could be gleaned from the reasons for leaving noted on the workers' files. This method was used by Paride Rugafiori in his study of Ansaldo, and he argues that a series of factors contributed to the high turnover in that factory, such as political sackings during the *biennio rosso*,[35] military service, mass redundancies in order to rehire the workers on a lower grade, movements of workers from one plant to another, and deliberate management strategies to divide the work-force. Rugafiori does, however, admit that the question is far from simple and that the sources are 'compiled by the factory itself, and it is certain that the truth is far more complex than appears from a simple examination of the reasons for leaving noted by the firm'.[36]

At Magneti, the problems with this approach are even greater since only one in five of the sample files gave reasons for the worker's departure from the firm and the information refers almost exclusively to the 1930s (see Table G, Appendix 2). Moreover, the quality of the information itself is highly problematic. The reasons noted are often vague and general (such as 'leaving of his/her own will') and too little information is recorded, so that an entry like 'maternity' fails to clarify whether the woman left voluntarily or was sacked as a result of time she took off because of giving birth. A more fruitful approach to understanding the dynamics of labour mobility lies in looking at which type of workers tended to be most stable. In other words, did the level of turnover vary according to sex, grade, and whether the workers were rural or urban?

Unlike in the Marghera factories, there seems to have been little significant difference between the turnover of rural and urban resident workers (see Table K, Appendix 2). This suggests that the Magneti workers had truly broken with the seasons of the agricultural year, a hypothesis which appears confirmed by a consideration of the monthly patterns of recruitment and leaving. Admittedly Table L in Appendix 2 does seem to show a seasonal pattern of recruitment, particularly for women, with

[35] The 'red two years'. This is the name given to the years 1919 and 1920 when the Socialists were at their strongest and, some would argue, Italy was on the brink of a revolution. See M. N. Clark, *Antonio Gramsci and the Revolution that Failed* (New Haven, Conn., 1977); P. Spriano, *L'occupazione delle fabbriche: Settembre 1920* (Turin, 1964); G. A. Williams, *Proletarian Order: Antonio Gramsci and the Origins of Communism in Italy 1911–1921* (London, 1975).

[36] Rugafiori, *Uomini, macchine*, 99.

the number of workers starting work rising in the second half of the year, but this trend may be more apparent than real as it is exaggerated by the sharp rise in the number of women starting work in the autumn of 1935 with the outbreak of the Ethiopian War. It may also partly be a result of seasonal variations in the demand for certain Magneti products such as radios. Furthermore, no significant trend in the pattern of leaving work can be discerned, so this wave of recruitment towards the end of the year does not appear to have been followed by an annual seasonal mass exodus from the firm (see Table M, Appendix 2). This somewhat ambiguous evidence does make it doubtful that any noteworthy link existed between the rhythms of the agricultural year and the employment patterns of these workers. It seems safe to suggest that the vast majority of Magneti workers no longer saw themselves primarily as peasants, giving priority to planting and harvest over industrial work.

One feature worth noting, nevertheless, is the slightly higher turnover of workers of both sexes from the city of Milan. In the case of the women this is not necessarily significant since the numbers are so small, but a good proportion of the men came from Milan and over half of these remained in the firm for less than one year. This is probably due to the fact that in the city there was more choice of employment and the urban dwellers were able, to a certain extent, to pick and choose between jobs.

More significant than the pattern of residence were the links between the rate of turnover and the grades of workers (see Table N, Appendix 2). Amongst the male workers, labour mobility tended to be highest amongst the unskilled and the apprentices, whereas the more skilled stayed for longer periods. The high turnover for the unskilled may reflect the fact that these were the most replaceable workers from the firm's point of view. Moreover, some of them were doing the more unhealthy jobs, such as foundry work,[37] and may have preferred to leave if they could find a cleaner job. The high turnover amongst the apprentices was undoubtedly at least in part due to military service.

But skill categories and residence patterns shed relatively little

[37] According to Pietro, an ex-worker I interviewed who had worked in the B Plant all the time he was in the firm, the turnover there was fairly high because of the potential health dangers of the type of work.

light on this matter in comparison with the sex of workers. Indeed one of the most striking features of labour mobility at Magneti is the extent to which it varied between the sexes. Looking at the male workers alone the turnover begins to look not unlike that found at Ilva, one of the factories studied by the Marghera group (although still far more stable than other firms such as Montecatini). In comparison, the female work-force seems to have been extremely stable: 28.3 per cent of the men stayed one year or less compared with only 18 per cent of the women, and 70.3 per cent of the women stayed in the firm for over two years.

My findings with regard to the importance of the sex of the workers to understanding labour mobility appear confirmed by Stefania Licini's study of labour mobility at Ercole Marelli,[38] Magneti's 'parent' firm located next to the Magneti A plant in Sesto. As Table 5.9 shows, here too the women (who were mostly winders and constituted a much smaller percentage of the labour force than at Magneti) had a significantly lower rate of turnover than the men. Only 25.4 per cent of the Ercole Marelli women left after less than a year compared with 36.6 per cent of the men. Furthermore, 37.7 per cent of the women stayed over five years compared with only 27.4 per cent of the men.

Licini argues that this phenomenon might be due to a series of factors, including the scarcity of opportunities elsewhere for women who had trained as winders and the possibility that women's upbringing made them prefer stability and continuity. All this is highly plausible. However, her subsequent analysis is weakened by her unwillingness to take gender seriously as a category: 'The possibility remains . . . that the pattern of stability can be explained, yet again, by geographical variables rather than others.'[39] In other words, since in her survey the Sesto born (she has no residency addresses) are more likely to stay longer in the factory, she conjectures that women may stay longer less because of the fact that they are women than the fact that they originate from near the factory. But it is also possible (and indeed desirable) to invert her argument and assert that the most stable sections of the work-force were both female and

[38] Licini, 'Operai a Sesto San Giovanni'.

[39] Ibid. 184.

Table 5.9. Labour mobility at Ercole Marelli

Length of employment	Percentage		
	Women	Men	Total
Up to 1 year	25.4	36.6	34.2
1–2 years	13.9	17.3	16.6
2–5 years	23.0	18.7	19.6
Over 5 years	37.7	27.4	29.6
Total as percentage of work-force	21.6	78.4	

Source: S. Licini, 'Operai a Sesto San Giovanni: composizione e mobilità della forza lavoro alla Ercole Marelli (1929–1945)', *Storia in Lombardia*, 3 (1987), 181.

resident near to the factory but the key to understanding their stability lies in their gender and not in where they lived. Women's very specific experience of factory life in this period can only with difficulty be subsumed into other categories such as 'locally resident'. Unfortunately Licini's tables are not all broken down by sex and it is therefore not possible to thoroughly check her hypotheses on these issues.

A clear focus on the sex composition of the work-force is crucial to understanding the pattern of labour mobility and there are, in fact, a number of highly convincing reasons why women's rate of turnover differed so much from men's at Magneti. Undoubtedly some of the higher turnover for men can be ascribed to military service, particularly in the 1930s when it seems to have accelerated to a certain extent (see Table O, Appendix 2),[40] but this is an insufficient explanation given the

[40] This modification of the situation in the late 1930s may, however, be to a certain extent illusory and in reality more due to the nature of the archive itself, since the files are loose and not in bound books. I was granted access to the files for the inter-war and Second World War periods, and it is not impossible that some of the files belonging to workers who stayed in the factory long after the Second World War were classified in another series. Unfortunately the archivist was neither able to prove nor disprove this hypothesis, since other files or the relevant *libri di matricola* were not found. If it is in fact true that I did not look at all the files for all the workers who started work before 10 June 1940, then, far from negating my previous arguments, this would seem to indicate that labour mobility was even lower than has so far been suggested, since it would be precisely the long-stay workers whose files would be reclassified.

extent of this phenomenon. Far more important was the labour market in Sesto and the way in which this varied according to sex. As has been argued above when looking at where the workers lived, the local labour market for men and women was very different. Women had less opportunity to find industrial or other adequately paid employment than men, and once they had found work at Magneti its pay and conditions may have been enough to encourage many of them to stay on as long as possible.[41] The existence of the nursery further enhanced their ability to continue working after the birth of at least their first child. Furthermore, as later chapters will explore more fully, men's and women's experience of work and family life were highly gender specific, leading them to view work and factory in different ways.

Male workers, although less transitory than those in some other firms which have been studied, still show a fairly rapid turnover and many were not in the factory for long. Nevertheless, in the context of highly industrialized Sesto and Milan their apparently casual relationship to factory work may make them no less clearly identified industrial workers, since many probably moved on, not to other types of work, but in search of promotion. Male engineering workers could seek employment in a number of other local firms and for some it was undoubtedly more attractive to move around from job to job.

A 'MODERN' WORK-FORCE FOR A 'MODERN' FACTORY

The work-force of Sesto's most 'modern' factory was, therefore, somewhat different from other nearby engineering firms such as Breda (which employed many male, adult, skilled workers). At Magneti about half the workers were female, and most were fairly young and unmarried and with no more than primary education. Furthermore, only a small number of recruits were classified as skilled.

Not only in Sesto but in the wider context of inter-war Italian industry this factory must be seen as somewhat unusual, both in terms of its commitment to Americanization and as regards the

[41] I am, of course, not rashly trying to argue that women in general tended to be less mobile as factory workers than men. This may or may not be true, but more research is needed into such issues.

stability of its work-force. It even might be argued that this was one of the first examples of a 'modern industrial work-force' in Italy, fairly stable and employed in a highly mechanized and 'scientifically organized' factory. The relatively low turnover varied to a certain extent according to skill categories, but the two factors which are most important to explain this pattern were management strategy and the large numbers of women in the labour force. Women remained in the firm much longer than men, a fact which is, at least to some degree, explicable in terms of the differing labour markets for the two sexes in the Sesto area.

As in stability of employment, in most other aspects considered in this chapter, from the workers' level of education to their place of residence, gender, more than any other category, proved a fundamental divide. The workers were all fairly young but the women were generally younger. Women, more than men, tended to be unmarried and childless. They also had had less education. Although as likely to be born in other parts of Italy as the men, reflecting the pattern of migration in family groups, many of the women lived nearer to the factory than their male counterparts. The male labour force was far more diverse with regard to factors such as age, skill, marital status, and education. Men travelled from further away to work and tended to be more transitory, moving from job to job in the more varied range of employment choices open to them locally. The female recruits, on the other hand, were mostly (although not all) untrained, poorly educated, youngish, and unmarried. They lived close to the factory and, once they had secured employment in this firm, they hung on to their jobs for as long as possible.

These statistics suggest, therefore, that the women represented a fairly stable and potentially docile group of employees for the firm, their jobs being so precious to them and the likelihood of finding comparable employment elsewhere so limited. Yet statistics can never give more than a broad picture. The results presented in this chapter must, therefore, be seen as no more than a preliminary and general profile of the work-force. Other, more qualitative sources can shed greater light on the issues thrown up by statistical investigation and can enable, for example, a closer examination of the nature of the work done by men and women in this firm.

6

'A Woman is a Perfect Machine': Gender and Work Roles

IN the early days of Italian industrialization women had not formed a large percentage of the labour force in the engineering sector, which had mainly employed men doing either highly skilled or heavy work. In the new field of light engineering, however, women represented a cheap, plentiful labour supply for the repetitive tasks of mass production. Employment in this sector was, therefore, a significant innovation in the history of women's work. Female employment in this sector should not, however, be simply regarded as the classic form of 'dilution' where cheap labour is used to pull down wages levels and standards of skill, since, as Veronica Beechey has argued, the employment of women in light engineering in the inter-war period was not, on the whole, the result of 'deskilling' within that sector but generally due to 'deskilling' within similar industries during the First World War.[1]

Moreover, in the case of Magneti Marelli, the exact nature of the relationship between an influx of female labour and progressive 'deskilling' is far from clear. During the inter-war period, the simplification and redefinition of work methods continued apace as production methods were continually modified. But this 'Americanization' does not appear to have been accompanied by a dramatic increase in the proportion of female recruits (see Table A, Appendix 2), although, given the greater labour mobility amongst the male employees, a steady recruitment of women would mean an overall gradual rise in the actual percentage of female employees. Unfortunately the figures for the 1920s are so sparse as to make comparison over time difficult. There were some examples of women taking over previously 'male' jobs (such as the replacement of some foremen

[1] V. Beechey, 'The Sexual Division of Labour and the Labour Process: A Critical Assessment of Braverman', in S. Wood (ed.), *The Degradation of Work? Skill, Deskilling and the Labour Process* (London, 1983).

by female *maestre* described below), but there is insufficient evidence to prove whether this was a process of significant proportions.

What does emerge from oral testimonies (inevitably referring of course largely to the 1930s and the war period, since few testimonies could be collected from those who began work in the early years of the firm's history) is that, although both women and men were employed at Magneti in a variety of different shops, on the whole work roles were strictly gender defined. Only in a few limited areas of the factory's work was it possible to find men and women working side by side doing broadly the same work. Furthermore, the female employees were virtually all confined to the A plant and then the N where the work was eventually transferred, whereas the B and C plants, where the work was mostly quite heavy, remained predominantly male.

Some of the women I interviewed did mention times when they would themselves have to take over men's work. If, for example, they were in a mixed pay team (not in fact a particularly common situation) workers might have to help each other in order to prevent the bonus rates from dropping:

> sometimes I even used to use a file, when it was close to Christmas or Easter, so that we could get our bonus payment. I even used to use a file . . . Working with a file was another type of work. It was men's work. I've even worked on drills. I've done all sorts of jobs. (Gina)

Cases such as these, however, seem to have been rare and were really the exception that proved the rule. Women only used files (and did other types of 'male' jobs), as Gina went on to explain, in specific circumstances. In other words this was only done 'to keep the bonus up . . . because the bonus was a team one'. Thus the changeover was only temporary and the gender definition of work roles remained intact.

WOMEN'S WORK

The files of the individual workers from the personnel archive of the factory reveal that female employees did a range of jobs within the firm. As Appendix 5 shows, apart from the assembly line workers, quality control inspectors, and those generally described as apprentices or auxiliaries, women were to be found working in the canteen and the stockrooms, operating milling machines, welding, winding, shearing, drilling, varnishing, and so on. On the other hand, this table also reveals how much narrower was the range of female jobs compared to those done by men, and just how many of the women, apart from the winders, seem to have been employed with the most general job descriptions. Indeed, 100 out of the 296 in the female sample were referred to simply as auxiliaries and a further 30 as 'workers' or 'general workers'.

In some factory shops there were both male and female workers, but their roles were usually different and the men tended to do a greater variety of different types of jobs. Some of the women were employed on the new automatic machine tools, at times doing work which in more 'traditionally' organized firms was reserved for highly skilled craftsmen. The majority of these machine tools were, however, still operated by men. One such shop where many women were to be found was the Shearing Shop. This was probably the most dangerous part of the factory where women worked, and a number of interviewees described it as a very frightening place. The women there operated the shearing machines and, in spite of the firm's health and safety policy, risked losing fingers or even a whole hand. This shop was mixed, with men and women doing substantially the same work, although the women were almost invariably given smaller machines.

A good proportion of the women worked in the assembly shops, and it was here that most of the women classified as auxiliaries or general workers were to be found. They worked on the assembly lines where each process was broken down into small, simple, repetitive tasks. The assembly was done in lines and increasingly on the new conveyor belts which were gradually installed in all the assembly shops. In these shops women's work was not physically demanding but required speed, precision, and concentration.

The Winding Shop was completely female except for a handful of male workers who did certain tasks such as portering and electrical repairs.[2] After the Second World War the work of

[2] According to one man I interviewed who had worked in the Winding Shop (Luigi B.) there were 9 men there by the late 1930s. None did winding.

this shop was divided up so that there were winders on most of the production lines, reflecting the different layout of the new N Plant in Crescenzago, but prior to this all the winding was concentrated in one shop in the A. No large machinery was used, only small machines to make the coils. The actual winding process was still done manually during most of the inter-war period.[3] Of all the jobs women did at Magneti it was probably the most demanding. Pierina describes it as 'quite heavy work. I had calluses all over my hands. You had to cut the copper wire to make the coils for engines using the shears. I had a callus this big!' Furthermore, a winder using a manual machine could do her job properly only after about five or six months. This was because manual winding was not particularly easy to learn: 'winding is very precise work. If you aren't accurate it's no good. It has to be accurate because it's the coils which make engines work' (Pierina).

The only real promotion open to women (apart from moving up to clerical work, which was only possible for those who had the right educational qualifications, or becoming a quality control inspector which represented only a tiny step up) lay in becoming a *maestra* (overlooker), like two of the women I interviewed from the Radio Assembly Shop. Frequently the introduction of American scientific management methods could lead to a deskilling of the work of the foreman. In the case of Magneti, one aspect of this was the takeover of some of the previously male jobs by women. As Orsola recalled, when she first entered the factory 'all the foremen were men. But after that they started to use women as they didn't have to pay them as much and it kept their costs down.' Their job consisted, to a

[3] Up until the mid-1930s all the winding seems to have been done manually. In 1935, however, a number of automatic and semi-automatic machines were installed, so that the work of some of the winders was effectively deskilled. See 'Vent'anni di lavoro nella FIMM', *SB* 1 (1937), 57. By 1937 the shop had 67 'winding machines of various types' and 68 semi-automatic winding machines. (See report by A. Spadaccino, Osservatore 203E, notizario 38 15 MI, ditta Fabbrica Italiana Magneti Marelli, Apr. 1937, ACS, OI, b. 22, f. 11, 'Impianti e macchinari'.) But it was in 1939 that a new type of automatic machine was introduced for some of the winding work, which meant that one woman could do work which previously had required a number of workers. (See 'Produzione in grande serie degli avvolgimenti per radio ricevitori', *SB* 1 (1939).) Even after these technological improvements, however, many of the winders continued to operate manual machines for much of the work.

great extent, in discipline and supervision. But the overlookers I interviewed did, nevertheless, display a certain amount of knowledge of the technicalities of production and were not simply figures of authority, since their work included assembling a sample of each new product that the line was to produce.

MEN'S WORK

Men, in contrast to women, had a long history of work in this sector. They dominated the work-forces of Sesto's other engineering plants such as Breda and Falck, and at Magneti, too, they were important. Although often working in the same shops as women, their role contrasted in many respects with the pattern of female employment. Their work was better paid, offered far greater chances of promotion, and tended to be more varied and interesting.

Men were employed in a whole series of different jobs at Magneti. From the new recruits' files it appears that many of the men worked as turners, radio engineers, assembly workers, mechanics, guards, smiths, electricians, inspectors, toolmakers, pug mill operators, fitters, and auxiliaries, and, in addition, smaller numbers of men did a whole range of other types of jobs. Men, therefore, could be found almost everywhere in the factory, in a wide variety of work roles. A few examples do, however, serve to indicate how most of these had been classified as suitable for men. At one end of the scale—the unskilled—men tended to be given heavier or dirtier tasks than women.[4] To this extent, women could be seen as getting a better deal than the men (apart from the level of pay). On the other hand, apart from winding, anything seen as 'skilled' (and better paid) was considered a male job. These more prestigious jobs were retained in the male sphere by a combination of tradition,

[4] The idea that these were somehow inherently 'male jobs' was reinforced by Fascist legislation. The list of jobs which women under 21 were banned from doing under the law of April 1934 (n. 653) included some which were relevant to light engineering, such as sandblasting and the handling of specific metals and alloys. Even before this law was passed, however, few, if any, women seem to have done this type of work at Magneti. (The regulations listing the actual jobs affected by this law were not published until 1936.)

which the management seem to have respected on the whole, and a monopoly of the necessary training and education needed to aspire to these roles.

Despite (or maybe because of) the Americanized management methods, a number of male, highly skilled workers were still necessary, such as the élite toolmaker-fitters and fitters. Some of the fitters worked in the mixed shops, whilst others were located in exclusive male preserves such as the Special Testing Room, or the Toolroom. Some were extremely mobile, working in a whole variety of different locations in the course of a single week:

> We used to get to move around . . . Because the shops where mass production was carried out were scattered about. And we went to repair machines. Because of this we had the possibility, we had special badges. We used to put a metal token here, a badge, and with that you could move around, the guards didn't stop you because you had the token. It was a kind of permit. And you could go from shop to shop mending machinery. (Giglio)

The fitters and the other skilled craftsmen commanded good wages and a respected position within the firm. Their work was often varied and interesting and as craft workers they generally had far more control over the work process and pace of work than many other shop-floor workers. Indeed, as one male ex-worker commented, 'it was every boy's dream to get taken on in the Toolroom' (Eugenio).[5]

Unlike women, men could aspire to rise in the firm, and although many male workers already had had more education than women, they could in addition top this up through the *Vedette* school. For some, this could be the beginning of a veritable career:

> after finishing the Magneti *Vedette* course, some boys could specialize and become good skilled craftsmen. Others could even go into administrative work if they weren't particularly good at manual work,

[5] This extract is taken from one of a series of oral testimonies collected by Luisa Finocchi for the book edited by L. Ganapini, '. . . *che tempi, però erano bei tempi* . . .': *la commissione interna della Magneti Marelli nella memoria dei suoi protagonisti* (Milan, 1986). I am indebted to her and the Milan CERISS (Centro di Ricerche e di Studi Sindacali) for providing me with copies of some of the transcripts.

or get a technical job like I did in the Technical Office. There were *Vedette* who became heads of shop or section heads. (Dino)

Only the most unusual woman could aspire to such heights. With such prospects, even a tough initial position could be seen as bearable for male workers if it could lead on to future advancement. As Eugenio noted, recalling his situation as an ambitious young man:

My aim was to learn a trade, but I was put into one of the worst shops in the factory, you could say, because of the atmosphere created by this type of work: it was the Automatic Machine Shop . . . But I liked it because I wanted to become a mechanic.

Threatened with redundancy from this shop he was very clear about where he wanted to go next and pleaded with the head of the firm's school:

I remember that I was crying with rage because I said 'I can learn a trade there' whereas in other parts of the factory it was all mass production and there were no trades, except in the Toolroom, because mass production had already started then. (Eugenio)

His fears were well founded. Far from all the Magneti men had skilled positions. American production methods meant that for men as well as women this factory contained many other less desirable jobs, with limited career prospects. Even some of the foremen's work had been greatly deskilled. In contrast to the 'traditional' figure of the highly skilled foreman who trained his own apprentices in the complexities of his trade, at Magneti the role of many (although not all) of the foremen was limited to keeping discipline and ensuring the supply of materials. Much of what might be seen as traditional foremen's work was now divided up amongst the new 'agents of Taylorism' such as the timekeepers.

Many other men, the 'unskilled', were trapped in what could only be seen as dead-end jobs, doing the heaviest tasks such as carrying materials or cleaning up. Virtually all the workers in the unhealthy, dirty atmosphere of the B Plant, for example, were male. Their work was arduous and could be dangerous. Others, the new 'semi-skilled', worked automatic machines and a number of the men were employed as guards or quality control inspectors. On the other hand, not many male workers were placed on the assembly line itself, except for the *Vedette* boys gaining practical experience in the various shops of the factory, for assembly line work was largely women's work.

CLERICAL WORK

Census statistics suggest that the number of Italian women employed in white-collar employment was rising in the early years of the twentieth century.[6] The clerical staff at Magneti Marelli appear to have been no exception to this. Although accurate figures on the gender composition of this firm's white-collar work-force are hard to come by, the evidence that does exist indicates that throughout this period the percentage of female clerks did grow. A communist report in 1926 recorded Magneti as having 159 clerical workers of whom only 38 were female.[7] This percentage rose steadily, particularly in the Second World War, and by the end of 1942, according to a secret government report, of 847 clerical workers in the A Plant, 268 were female.[8]

Yet this increase was always, as ex-clerk Maria B. noted, 'within certain limits for the women'. In other words for clerical workers, like manual workers, work roles were gender specific, and women's role was seen as subordinate. Female clerks were not, as in Victorian England, physically separated to preserve their modesty,[9] but, although they worked in the same rooms as the men, their tasks there were different and 'the men got the best jobs. Naturally the women had to help the men prepare the work' (Maria B.). Women's work was largely confined to the more mechanized and repetitive aspects of office work. A few were shorthand typists, but the majority did more menial tasks such as the filling in of pay slips using special machinery, copy typing, or filing. The largest number of women seems to have always been found in the pay department, which constantly

[6] See e.g. G. Tagliacarne, 'L'occupazione delle donne nelle fabbriche e negli uffici durante gli ultimi cinquant'anni e la diminuzione delle nascite', *Giornale degli economisti e rivista di statistica* (Dec. 1934), 6.

[7] 'Dalla Magneti Marelli di Sesto San Giovanni', *Unità* (31 Mar. 1926), 3.

[8] ACS, OI, b. 22, sf. 'FIMM: personale in forza al 31. 12. 1942'.

[9] See e.g. M. Zimmick, 'Clerical Work for Women, 1850–1914', in A. John (ed.), *Unequal Opportunities* (London, 1986).

generated a lot of paperwork due to the complex bonus pay system. Within each clerical department it was always men who were in control, commanding the best and most prestigious positions as office managers or accountants. Women may have filled in the pay slips but it was male clerks who carried out the 'technical' role of actually timing the workers and making the calculations. Indeed once work was seen as in any sense 'intellectual', it was defined as automatically for men. As Isa remembered, the men 'did the more conceptual work, let's say the more technical work more than anything'. As she noted, women simply were not trained to do this type of work:

> Maybe (it was) because the men had more education, I'm sure it was. Because in engineering they had been to school longer. We women, on the other hand, had probably only got as far as the commercial classes at the very most. (Isa)

But it was more than simply the impact of limited educational opportunities which produced this segregation of roles, since it minimized potential conflict over the expansion of women into the higher-status clerical work and ensured that male privilege and status remained intact.

WAR

Wars and rearmament were a key to Magneti's growth. As Table A in Appendix 2 shows, they also led to an increase in the number of women recruited. The influx of female recruits in the 1935–6 period and in 1940 indicates that in the build-up to the Ethiopian War and Second World War, the firm was increasing the proportion of women as men were called up to fight. Despite this it appears that the pattern of gender roles in the work-place was affected only to a limited degree. The statistical material presented in the last chapter is a poor yardstick for this, since the sample ended with the outbreak of the World War. Evidence from oral testimonies, however, suggests that many male engineering workers were exempted from military service and retained in the firm. Even the government inspectors investigating the various war industries seem to have shared the view that there was a limit to the extent to which women's role could be expanded. In 1939 a secret government report on the Magneti B plant concluded with the assertion that 'the eventual possibilities of using substitute female or child labour in case of emergency—[are] none, given the type of work'.[10]

It is fairly certain none the less that some encroachment onto hitherto male territory did take place. For example, one male ex-worker remembers that some new women were taken on because it was cost-effective:

They gave them some of the work because the pay was different, it was lower, they used women, they exploited them more. Do you understand? Not only that, but when you think that when the Time-Keeping Department fixed the bonus rates, if the work . . . was done by women [it was] at women's pay rates. If on the other hand men did it, that piece of work gave a bonus according to the men's pay. So it became more expensive. (Gaetano)

Evidence from the factory archive also backs up this idea. In the 1938–41 period numerous references from the personnel files indicate that a lot of regrading was taking place and significant numbers of women were moved up to the A grade as they started work on the Boley lathe or were transferred to the varnishing shop to do certain, often previously male-designated jobs.[11] It does seem likely, therefore, that in the 'grey areas' where male and female roles were similar some takeover of men's jobs did occur. On the whole, however, the so-called 'process of dilution' did not take place on a great scale. In contrast to other countries such as Britain, where during the Second World War at least some women trained in crafts like bricklaying, fitting, and welding, there is no evidence at Magneti Marelli of women being given the chance to acquire what were seen as 'traditionally male' skills. Women do not appear to have been allowed to threaten the male monopoly of better-paid, more interesting work and learn trades such as

[10] Report by Giuseppe Fedelini—Ufficiale Osservatore 203E, June 1939, ACS, OI, b. 22, f. 1.

[11] Many of these workers' files in the APMM contain the following memorandum: 'Whenever these workers are put on to other types of work in other shops they should lose their hourly pay of L. 1,24.' Thus this change of status was purely temporary and the women only received the higher pay whilst they were needed on this type of work. As soon as they were no longer needed, they were regraded back down to the B category.

fitting. 'Skilled work', apart from winding, remained a closely guarded male preserve despite the pressures of wartime production.

GENDER CONFLICT?

On the question of gender conflict in the work-place the surviving evidence is somewhat contradictory. On the one hand the entry of women into the male-dominated world of engineering was potentially a source of resentment for the men, and some limited evidence indicates that this could emerge at times. On the other hand the bulk of the oral testimonies suggest that this was not a major issue.

One example of tension and resentment is an article written in 1934 by Mario Palma, Magneti *squadrista* and activist from the Fascist unions. Palma's article appeared in the newspaper of the Fascist unions, *Il lavoro fascista*. It is worth examining, even if it is not possible to know how widespread similar views were, since it does show the type of attitudes which could surface in a crisis when men felt threatened. The article was concerned with the problem of male unemployment in Sesto San Giovanni. According to Palma this unemployment was caused

> partly by the modern mechanical systems of standardized work methods, but also to a great extent it is due to the far too extensive use of female labour: even in engineering firms women do jobs that were previously carried out by men: big shearing machines, automatic lathes, grinding machines, drills, milling machines, and nickel baths are all worked by women, thereby damaging male employment prospects.[12]

As a good Fascist, he worried that 'With all this not only will we fail to achieve the demographic increase that we want but it will also lead to a progressive weakening of the race.' His solution was for the bosses to stick by the male workers, since he (most implausibly) asserted that it is only because of the male workers that these firms had survived and prospered:

[12] M. Palma, 'Maestranze femminili in pantaloni, grembiuli e maschere', *Il lavoro fascista* (1 Sept. 1934). It is worth noting that Palma, although a Magneti employee himself, was not necessarily referring to this firm here. It is intended as a general comment on Sesto's factories.

Well, now that the lean season has arrived, maybe we should remind those gentlemen who set up business in Sesto twenty years ago, that their firms have grown because of the work of these male workers and we should ask them to shut their eyes for a moment and try to visualize the army of unemployed men and if they have the slightest conscience they should . . . send home to be housewives any women whose husbands are employed and generally get rid of any other female labour with the sole exception of any genuinely needy woman.[13]

According to Palma not only financial issues were at stake here, but the threat to masculine pride, rendering the situation intolerable:

How many fathers must blush before their own little daughters who are out at work to try somehow to save even their fathers from starvation? How many wives are there who return to find their husbands doing the cooking, trying to make a plate of soup for their wives who are out all day working in a factory?[14]

This attitude was typical of much Fascist rhetoric about women's work. But just how widespread these kinds of views were at Magneti is less clear. Occasionally recollections of this type of misogynous attitude did surface in the oral testimonies I collected: 'certain women, in certain jobs, will agree to do men's work without a man's pay . . . (this) did create some ill feeling' (Costantino). However, the testimonies of the majority of the ex-workers I interviewed suggest that in general there was a striking absence of gender conflict over roles.[15] When the male interviewees were asked if they had felt threatened by the influx of women and whether they had felt afraid that their own jobs could be imperilled by this the most frequent answer was no. On the whole, men seem to have been singularly unworried by women's presence in the factory. Why then was such a potentially conflictual situation apparently so easily accepted by men? This question can be answered in a number of ways.

Firstly, the concept of a 'family wage' with one male

[13] Ibid.

[14] Ibid.

[15] Clearly, the fact that I, the interviewer, am female may have influenced the way in which these male workers chose to remember their own attitudes towards women's work at the time. Despite this, this theme seems far too persistent in these testimonies to be purely a product of this factor.

breadwinner was illusory for many households, and working-class families usually expected most of the members to contribute. Wives without small children (and often even those with one or two small children) generally worked, and certainly daughters above school age would only be kept at home if the father was exceptionally well paid, or alternatively to do housework and child care to free the mother for paid employment. Although in times of economic depression some may have agreed with the Palma article quoted above, most men accepted female wage-earning as necessary to the finances of their own household. Indeed, some of these men were themselves sons, brothers, or husbands of female Magneti workers.

Secondly, Magneti was an expanding concern and the influx of women did not usually lead to a man losing his job. This factor emerged in many testimonies. For example, one male worker asserted that there was no gender conflict over issues of 'dilution' because 'when there is plenty of work there's no question of resentment' (Gaetano). Another, when asked if the men were afraid women would rob them of their jobs, replied 'No! Because there was enough work for all. There was no lack of work' (Luigi B.). Apart from the period of the slump, employment prospects for male engineering workers, particularly if skilled, were generally fairly buoyant. This factor was significant in easing gender conflict, particularly from the rearmament for the Ethiopian War until the end of the Second World War.

Thirdly, women themselves failed to adopt what could be seen as a threatening attitude. When I asked one male ex-worker if any of the women aspired to the really élite male jobs such as becoming fitters, he replied:

> No! Not in those days, no! . . . In my opinion that wasn't even an issue. Far from it . . . Because a woman in those days, as far as I can tell, still felt like a nobody, she felt like she was just there to do her work and her housework . . . You know, I've been all over that factory, I moved around a lot in the production shops because of my job, and I never once found a woman who was as emancipated as, I won't even say like nowadays, but even with the emancipation of twenty years ago. (Ercole)

The fourth and perhaps most important factor to consider was the generally rigorous gender segregation described above. Men and women at Magneti were all engineering workers and at least superficially it could appear that women were encroaching on a traditionally male preserve, but the actual jobs that they did were carefully labelled 'men's work' or 'women's work'. Since many of the jobs that women did were new, and did not have a precedent in other older engineering firms where the work was heavier and organized differently, conflict was reduced. The following two replies by male ex-workers to the question of whether men felt threatened are typical:

No! Because there they gave women a certain job that only women could do and that was all. (Pierino)

No! Not in those days, no! Because men had their own specific jobs. Which in those days they said women weren't capable of doing. (Ercole)

In other words, so long as women only did 'their' jobs and did not attempt to get men's jobs, particularly the more desirable ones, the men accepted it. This management technique seems to have worked well, since both men and women appear to have accepted this definition of which tasks were fit for each sex. For example, one woman told me:

[Women's work in the Shearing Shop] was accepted because it was unlikely that a man would go onto such a small shearing machine. There were men on the shearing machines but only two, four, five machines. Heavy ones . . . [In another shop where mostly women worked they made] little things, really little things. That work was in baskets as small as this which might have even a thousand of each thing in them. Very tiny things they were. A man couldn't say, 'There are too many women!' because it was a type of work . . . (Isa)

There were numerous examples of this type of comment in the testimonies I collected. Giuseppina, for example, commented that in the Spark Plug Shop '[they were] almost all women, because all those jobs were women's work'. Luciana backed this view when she noted:

but there were hardly any men, we were almost all women in the shop, all women, even the overlooker was a woman. Higher up there was the male head of shop and a few foremen, the ones that were needed to assemble a machine, or to put something right or change a tool: maybe only a small tool, but you need a man for that job.

She reinforced this idea with the comment that 'A lathe is a man's machine because you have to fit the tools and change them and so on.'

Moreover, some of the testimonies of both women and men even suggested that women somehow benefited by being limited to 'women's jobs'. One ex-worker (Ercole) recounted his discomfort when women were given shearing machines that he considered too heavy for them. Another woman recalled:

> The assembly work was women's and the actual work on the frameworks was men's. Because they had to be turned in a lathe and put together, some of the pieces that had to be turned were quite heavy and they had to use those compressed air drills. Then there were women's jobs, such as the assembly of the coils or other things and they had to be welded. Women were in all the shops. Magneti Marelli was full of women. The majority of the Magneti workers were women because even the radio . . . You know, they were all nice little jobs. (Giuseppina)

MASCULINE AND FEMININE 'SPECIAL SKILLS'

Not only did both men and women seem to accept their differing roles but many went further and argued that each sex had its own specific contribution to make to the productive process. Indeed, in the oral testimonies I collected one recurrent theme is the frequent reiteration by many of the ex-Magneti employees that men and women were good at different things. Rarely did anyone face head on the issue that in general women were confined to jobs which were among the least desirable. On the one hand, men were generally thought to be good at 'intellectual' or physically heavy work. They alone were seen as capable of highly-skilled technical work or craftwork. Only men were believed to have the special abilities needed to become fitters or electricians. Women, on the other hand, were presented as possessing specific female skills of endurance, nimble-fingeredness, and patience and were thought to be good at any task that was delicate, repetitive, and monotonous.

For example, a female former clerical worker explained why women were chosen to do the winding:

they were very, very delicate processes, more women's than men's type of work . . . with all the bobbins, the tiny things needed for the injections and the bonding, minute things they were. Because of that it was more women's work. (Maria B.)

One of the ex-workers I interviewed took this theme even further:

the Shearing Shop was all women. It was quite a dangerous job. Well there something amazing happened. After the war they decided that women shouldn't work on the shearing machines any more. And look what happened! The output . . . Because women, in manufacturing you need three men to do as much work as one woman. A woman is like a machine . . . A woman knows how to synchronize her work, she really knows how to do it! When men replaced the women on the shearing machines, there were more accidents and half the output. A woman is quite something! . . . In manufacturing a woman is like a machine, her work is so well coordinated and she works so hard . . . But in manufacturing it was men who were the failures, not the other way around. (Ercole)

This same worker argued that this swifter work was not due to women's greater fear of losing their jobs, but due to their own inherent skills in this type of work:

let's say that in that type of work a woman doesn't get tired . . . a woman . . . is a perfect machine in every respect. A woman is a perfect machine. If only someone could invent a machine as good as a woman . . . it would be a marvellous thing wouldn't it. It's not because women are afraid [that they work harder]. Women, my mother for example, are stronger than men. For them, a repetitive job isn't boring, they're made of different stuff than men. (Ercole)

MASCULINE PRIDE SAFEGUARDED

Although maybe not all of the men (nor the women in all probability) would have concurred with Ercole's view that one woman was worth three men, the majority of the work-force do seem to have agreed that men and women had different special skills at work. Each sex was thought to be good at certain gender-defined tasks and at these alone. Thus gender conflict was minimized and a rationale supplied for the strict definition and separation of male and female work roles. All this, coupled with the general expansion of the firm which made jobs plentiful, meant that the management was able to introduce profitable female labour into the hitherto predominantly male world of engineering with comparative ease. With the retention and protection of male superiority in the workplace, only during the period of economic depression did women's work appear to threaten masculine pride and job security. At these moments, however, misogynous Fascist rhetoric against women's work could surface.

On the whole, therefore, it appears that, for a series of reasons, men accepted, or at least failed to resent greatly, the large-scale employment of women in this firm. As far as the women themselves are concerned, it seems that they, as much as the men, accepted the rigid assignment of gender roles in the workplace. But this still leaves many questions about women's entry into engineering unanswered, not the least of which is the issue, only barely touched upon in this chapter, of how women themselves experienced work in this new alien world. How did women working in the most modern and male-dominated of industrial sectors, under a regime which was publicly hostile to female employment outside the home, condemning it as unpatriotic, view their own situation?

7

'I'd Have Felt Like a Thief if I'd Stayed at Home Doing Nothing': Women at Work

CERTAIN themes which emerged in the oral testimonies I collected proved particularly useful to understanding more about the female experience of work at Magneti Marelli.[1] Firstly, one recurring theme in many of the interviews with women was the question of relationships with other workers, in other words how they experienced the collective life of the great factory and the extent to which their experience in the work-place was primarily characterized by solidarity or conflict. The second area of noteworthy interest was the series of vivid accounts given of the virtually all-female Winding Shop. Finally, these interviews revealed a great deal about women's subjective attitudes to their work and the firm itself. They enabled an exploration of the issue of whether the female workers in Italy's foremost Fordist factory shared the work ethic found by Luisa Passerini[2] to have been prevalent in Turin's working-class community in this period or whether, conversely, these women had internalized the negative attitudes to women's work in waged labour that was an intrinsic part of Fascist ideology.

RELATIONSHIPS, FRIENDSHIPS, AND DIVISIONS ON THE SHOP-FLOOR

Many of the female interviewees remembered the companionship of the work-place as a positive aspect of their work experience at

[1] For an earlier version of some of this chapter see my article, 'Lavoro femminile, ideologia del lavoro e relazioni in fabbrica', in L. Ganapini (ed.), *'. . . che tempi, però erano bei tempi . . .': la commissione interna della Magneti Marelli nella memoria dei suoi protagonisti* (Milan, 1986).

[2] According to Luisa Passerini a deeply held work ethic was one of the 'bed-rock' assumptions of the working-class people she interviewed in Turin. See

Magneti, and spoke of friendships at work with nostalgia. Luciana, for example, said 'there was more togetherness, people helped each other more. People had better manners in those days.' Some of the women I interviewed remembered even the relationships between the overlookers and their work-teams as being good, and the women working under them did not see them simply as figures of authority representing the management. Carla commented that 'the overlookers were more scared of the management than any of us'. Germana, herself an ex-overlooker, recalled working with her team with nostalgia, saying, 'we used to have a laugh, we joked a lot, it wasn't that bad at all'. Giuseppina felt that the explanation for this lay in the fact that 'the overlookers had to work hard too, they didn't get a moment's rest', in contrast to the foremen, whom she described in the following manner:

> the foremen were sometimes men who'd risen up through their own hard work, but a lot of them were people who'd got promotion simply because they were bigmouths. A lot of them were much better at talking than working. That was the trouble with them.

She felt, on the other hand, that the female overlookers 'were more humane. After all, they were women. They were really understanding. Personally, I never had any problems with the overlookers.'

Even the relationship between local workers and those who commuted into Sesto from rural areas was remembered positively. Antifascist sources suggest that the willingness of the rural workers to accept worse conditions led to fear and resentment amongst the urban workers in this firm:

> [The rural resident workers] are mostly unskilled and fifteen per cent of them have a small piece of land which they cultivate in their free time. These workers, who fill out their wage packets with agricultural work, are not approved of by the majority of the workers because 'they accept starvation wages more readily'.[3]

Passerini, 'Work Ideology and Consensus under Italian Fascism', *History Workshop*, 8 (1979); Passerini, 'Fonti orali e storia della classe operaia in regime fascista', in A. Agosti and G. M. Bravo (eds.), *Storia del movimento operaio, del socialismo e delle lotte sociali in Piemonte* (Bari, 1980); Passerini, *Fascism and Popular Memory* (London, 1987).

[3] 'Dati e considerazioni sulle condizioni di lavoro degli operai della "Magneti Marelli" ', *Lo Stato Operaio* (15 June 1939).

But the women I interviewed contradicted this view. For example, Carla stressed that

in that respect there was no difference between us. Firstly, there wasn't any time for ill-feeling to develop because everyone tried to get on with her work and if we could we helped each other as well, especially the new recruits. It may seem impossible but we all got on well.

Maria G. said the same: 'we got on fine even if we spoke different dialects. We often had a good laugh because we couldn't understand each other, but we got on well.'

It seems, therefore, that among the women, even rural and urban women, friendships did develop and there was some degree of solidarity. However, relationships in reality were not as simple as these rosy memories suggest. It seems that the fact that these workers 'were women', as Giuseppina mentioned, did not automatically lead to harmony on the shop floor.[4] In fact, a number of the interviewees remembered the relationships amongst the women as marked by jealousy, and they contrasted this with the easier relationship they had with the men:

Working with men was much better, that's certain. I hope you never have to work only with women, it's terrible. There's so much jealousy. (Carla)

It was often easier to get along with the men than with the women. You know, in a group of women there's always one who puts on airs, thinks she's better than the others. That doesn't happen with men. (Bruna)

Men are more reasonable than women. It's easier to get on with men than with women because men are more easygoing. Women are more jealous about their work. (Luigia)

These frequent references to jealousy among the women workers raise important questions about these women's experience of factory life. It seems probable that such problems were caused, at least in part, by the fact that women were less prepared than men for the sort of relationships they would

[4] The testimonies of the women that I interviewed well illustrate Janet M. Bujra's assertion that it is not enough simply to assume that women have a natural 'sisterhood', but instead it is necessary to study more carefully the conditions which lead to solidarity or divisions among women. See J. M. Bujra, 'Female Solidarity and the Sexual Division of Labour', in P. Caplan and J. M. Bujra (eds.), *Women United, Women Divided, Crosscultural Perspectives on Female Solidarity* (London, 1978).

encounter in the world of work, being brought up to identify primarily with the home and family and having only had a restricted amount of contact with the wider world. This made them more insecure in the alien world of work, uncertain of their ability to keep up with the pace of work and suspicious of others who might work faster. Added to this, jealousy and friction were exacerbated by the particular organization of the work process in this firm and especially the bonus incentive pay system. As the women almost always worked in all-female groups, on the whole this meant that they were in competition with each other rather than with the men.

Practically all of the women were paid according to a bonus system. This was individually calculated in the Winding Shop and collective in most of the other shops where many women were to be found. As Chapter 4 showed, frequently workers attempted to keep the bonus rates up by slowing down the work pace, or even hiding finished work. These attempts to defy Taylorism obviously required a certain amount of solidarity, but the bonus rate system was also a source of conflict and jealousy. These problems existed whether the bonus system was collective or individual. Bruna remembered that 'when we worked on an individual bonus system everyone was always complaining, for example, if someone needed a certain tool when another woman was using it'. Furthermore, she said that 'some people worked slowly but there was always someone who went too fast and then everyone suffered'. With the collective bonus system things were no better: 'If anyone went too fast the others would shout at her "Why did you work so fast? You shouldn't have done!" ' (Luciana). Even if the workers tried to keep the pace of work at a reasonable level the collective bonus system was, in fact, an excellent method for ensuring that everyone worked hard, for if there was one slacker the whole team suffered. This intensified the atmosphere of suspicion and meant that, to a certain extent, discipline was imposed by the workers themselves. The reason for this was that 'if someone didn't want to work hard then she was profiting from the bonus that the others had worked for' (Bruna).

In addition, according to one of the women I interviewed, friction often arose from the fact that the older women were afraid of not being able to keep up with the young. In the unskilled or semi-skilled work done by women in the 'rationally organized' factory speed was often more important than experience, since the work could be quickly learned by a young girl, who, after a short training, was frequently able to work faster than a woman with years of experience in the firm:

Some of the women, especially the older ones who had been in the firm a long time, were worried when young girls started work . . . . It was because of the bonus system, you see. That's what caused the jealousy. You see they were afraid that when some young girl who could work fast turned up, then the bonus rates would be cut. You see what these older women were like! So when these young girls started work I really felt sorry for the poor things, because I went through that myself. Because the older women who worked there weren't really very good at their job. (Giuseppina)

THE WINDING SHOP

One part of the factory, the Winding Shop, was described at great length by virtually all of the female interviewees (and by many of the men too). The accounts of this shop afford a good deal of insight into both women's experience of work and the firm's attitude to its female labour force. This virtually all-female shop was ruled over with a rod of iron by the only female head of shop in the factory—the formidable Luigia Pedroni. One former clerical worker from Magneti gave this vivid description of her:

At Magneti there was one shop where only women worked and it was probably the shop which functioned best in the whole factory. The head of this shop was a woman. Signora Pedroni was like a character out of the 'far west', she wore nineteenth-century clothes, long skirts and her hair was pulled tightly into a bun in the middle of her head . . . Her shop was very tidy and organized. The workers had light blue overalls and when they came off their shift—at a different time from the rest of the factory so they didn't have to mix with the crowd—they looked as though they'd just come out of a boutique . . . The women worked with enthusiasm, it was a kind of work that needed a certain dexterity and quick reflexes.[5]

[5] 'S.S.G.M.' [interviewed by P. Crespi in the Convento delle Suore Collettine Francesi, June 1979], 'Le fabbriche di Sesto hanno vissuto drammi storici', in P. Crespi, *Capitale operaia: storie di vita raccolte tra le fabbriche di Sesto San Giovanni* (Milan, 1979).

The women I interviewed all had vivid memories of Luigia Pedroni (one even executed an imitation of her walk for me!) Bruna commented that 'she scared everybody. She was an impressive woman—you only had to look at her to know that.' Pierina summed her up as 'very strict. A serious woman, never frivolous and even beautiful too, but she really was very harsh.' For Giuseppina 'she was worse than a dictator, she wore a black overall with a high collar, never had a hair out of place and never wore a scrap of make-up. She was like a general!' Under this authoritarian figure, discipline was very strict. Relying on her own powers of intimidation, she dispensed with guards and 'patrolled the shop all day long just like a policeman' (Pierina), or watched over her work-force from a raised cabin. Her powerful air of authority was enough to keep order: 'we really had to work with her around. Everyone was afraid of her . . . you know even women who were usually really arrogant were like little lambs when they were in her shop' (Carla).

The shop was kept permanently locked and all other people, even members of the top levels of management, had to ring a doorbell to be let in. The winders worked to a different timetable from the rest of the factory since, as Maria V. said, 'she didn't want her workers to have any contact with the men, so they had that special shift and that was all'. Work began at seven in the morning and ended at half past three with only a half-hour lunch break which did not give the winders enough time to go to the factory canteen. At the workbench they had to dress and behave modestly. Maria G. remembered it as 'the most orderly shop in the whole factory, no short sleeves, no make-up, no untidyness, no eating'. Even in the stifling heat of summer, in spite of the fact that the windows had to be kept tightly shut to prevent draughts which could ruin the delicate work, long buttoned-down sleeves were compulsory.

Pedroni's separatist regime meant that the winders were effectively sealed off from the rest of the factory. Maria V., who started work in 1933, saw nothing more of the factory than her own workplace and the Personnel Department in the first ten years that she was in the Winding Shop. Few winders made use of the firm's leisure facilities, and only those who had contact with antifascists outside the work-place got involved in the political activity developing in the other shops of the factory during the period of the Resistance. In one instance the consequences of this isolation were tragic. Because the door was locked, first aid could not be brought in quickly enough to save the life of a woman burnt alive in an industrial accident:

> if the door had been open at least we could have got help. Because all it needed was for someone to go and fetch help. Instead by the time we all jumped out of the windows she was covered in flames. They threw blankets over her but it must have been a terrible way to die. (Carla)

Yet the most intriguing aspect of these descriptions of the Winding Shop was the attitude of the winders themselves to this rigid discipline. Some of the women did not have good memories of working there. One of the more politicized of the interviewees (she was a messenger in the Resistance and later an active member of the PSI) condemned Pedroni and said that 'not everyone liked her. I can't believe that anyone said that' (Carla). But even Carla's attitude was ambiguous, since when Pedroni was sacked at the end of the war she herself felt that the fact that Pedroni was a member of the Fascist Party was irrelevant: 'Well, she was a member, but so were lots of other people'. She also remembered, with grudging admiration, an incident when Pedroni defended her workers against the criticisms of an engineer. That other workers also felt critical of Pedroni is evident in the fact that in 1943 a delegation tried to get her sacked by appealing to Quintavalle, the managing director. Pierina, one of the protagonists of this episode, recounted the story:

> I went to see the Managing Director, things were like that you know around the 8th of September, and I wasn't the only one, there was a group of us, mostly young people, and we told him that Pedroni was a Fascist, and this and that, you know the sort of things that you do when you are young.

But their protest was in vain and the protestors, including Pierina herself, were sacked.

It does seem that some of the winders were far from happy working under Pedroni, but what is remarkable is the fact that a number of the women say that they actually liked the atmosphere of the Winding Shop. For example, Bruna said '. . . as far as I'm concerned she was a good woman, a good person'. Maria G. also expressed this view: 'Even though she was so strict, for myself, I liked it there, and so did some of my workmates.' She even went on to say:

> She protected our shop. It was a shop where no one had any cause for complaint. Even if there had been strikes, if a worker is respected by the firm and works hard, she has no reason to protest. Protests arise when the workers feel a need for them.

One reason why women felt good in this shop was that it represented a safe and 'respectable' environment. The notion of 'respectability' was very important to women (see Chapter 9) and even in factories where there were few men, such as textile mills, there seems to have been a certain amount of concern about the 'loosening of morals' amongst female workers.[6] In a firm like Magneti, where the work-force was much more mixed, these 'dangers' were increased. One solution was to create an all-female environment, safe from male encroachment.

Moreover, quite apart from the 'danger' that women might become 'sexually immoral', women workers themselves had to contend with the real problem of sexual harassment when working with men.[7] For example, one of the accusations specifically made against Umberto Quintavalle by the CLN after the Liberation was that:

> Ing. Umberto Quintavalle quite immorally used to give out very well paid jobs to persons who degraded themselves by marrying female clerical workers who were well known to have been his lovers.
>
> In the firm there were numerous men who had got into and kept

[6] e.g. the sexual mores of young female textile workers in nearby Legnano were considered important enough for a local policeman to send a long report about the situation to the Prefect of Milan. The author of this report was concerned about the girls of Robecchetto, where 'as in the other villages in the Legnano area, the majority of the industrial workers are women who work hard, earn their living, and like to go out and enjoy themselves. They generally like good company and for this reason they are often easily conquered by the more enterprising young men in the area.' He stressed that these were not prostitutes. 'In the Legnano cotton centres every girl lives by her own honest, hard work and if she does concede her favours to some young man it is not for financial gain.' Yet despite this, the report's opinion was that 'the morality of this area is far from good' and the reason for this state of affairs lay 'in industrialization and the sort of factory life these girls lead'. (Ten. Colonello Angelo Cercia to Prefetto di Milano, 6 Apr. 1929, ASM, Gab. Pref., cart. 225, f. 'Pubblica moralità'.)

[7] For a study of sexual harassment, see J. Lambertz, 'Sexual Harassment in the Nineteenth Century English Cotton Industry', *History Workshop*, 19 (1985).

themselves in good jobs by supplying the two owners with women from amongst the workers and clerks.[8]

Some of the interviewees mentioned incidents of harassment, usually with reference to foremen, heads of shops, or managers. Giuseppe noted that 'Before the war, that's for certain, those who had positions of power sometimes abused that power, especially with regard to the women.' Luigi, who worked in the Winding Shop, claims that, although the few who worked there were very respectful, in the other shops it was different. As he notes, some of the women knew how to retaliate: 'In the other shops . . . every now and then there would be a quarrel. The women would box the ears of some man who was misbehaving himself. Things like that happened . . . you know, they were only human.'

The Winding Shop represented a safe haven from this type of trouble. For women who had had little contact with the outside world, an all-female environment ensured them a greater sense of security in the alienating world of the large factory.[9] Maria G. described how difficult it was for her when the winding work was transferred to the N plant in Crescenzago and divided up among various mixed shops:

> I found it totally different, having to leave my own shop and fit in with everybody there. It was as different as night and day. In Pedroni's shop our lives were regulated, we did our job, we earned our money and that was that. In the other shop we had to adapt to the outside world.

According to some of the interviewees, another reason that women liked working in this particular shop was that the pay was relatively good. The winders do seem to have been paid better than most of the other women in the factory, and Maria G. even went so far as to say (with some exaggeration) that 'the

[8] CLN FIMM comunicazione 033, ACLNL, b. 4, f. 13, p. 3.

[9] This method of rendering more acceptable women's entry into factory life by separating male and female spheres was not uncommon. Another example was the Courtauld's rayon plant in Coventry, where work roles and even the canteen were segregated by sex, and a special nurse (Nurse Gaskin, who appears to have been a very similar figure to Luigia Pedroni) was employed to supervise the female workers' morality as well as their health. See J. Castle, 'Factory Work for Women', in B. Lancaster and T. Mason (eds.), *Life and Labour in a Twentieth Century City: The Experience of Coventry* (Coventry, 1986).

women were paid like men . . . Magneti Marelli was a firm where you could really earn well.' Maria G. was not the only one of this opinion: 'In spite of it all, everybody wanted to go and work in that shop because they always took home loads of money, more than in the other shops' (Bruna). Furthermore, work conditions were good. The women worked sitting down and the shop was clean. Admittedly there were bakelite fumes in the shop which might have endangered health, but, to guard against this risk, there were special fifteen-minute rest breaks morning and afternoon when the workers were allowed to go out into the fresh air.[10]

The actual work of winding itself was a source of satisfaction. Giuseppina commented that 'the work was lovely, it was winding!' and Pierina said 'I really liked winding, it was a kind of work that I really liked. Even though it was quite heavy I was always happy to do it.' Indeed the very fact that it required some skill was one of the reasons the work was seen as enjoyable. As Maria V. remembered with nostalgia, 'I really loved it, I loved work, I even loved the work itself. It was quite hard, but when you'd finished you really felt satisfied.'

To a certain extent the women accepted the discipline of the Winding Shop, and although they recognized that some of the rules were severe they felt that Pedroni was fair towards them and defended their interests:

> I always liked it there because we were respected and earned well. It was a nice place to work. When Pedroni was there, if we didn't earn a hundred per cent, she refused to deliver the engines because she said that her women worked hard and she expected them to earn well. She used to sit perched high above us watching over everything we did and she didn't let us get away with a thing, but she made sure we were well paid and respected. (Maria G.)

According to another interviewee, Pedroni even defended her workers against the time study engineer:

[10] These rest breaks are yet another example of the firm's 'rational' approach to health and safety described in Ch. 3. As the factory newspaper was keen to stress, this initiative raised productivity: 'following American examples, in one shop where many women work, we have introduced a quarter-hour rest period every two hours. The result of this has been that overall productivity has risen by 5% and the daily output charts reveal that there is a sharp rise in productivity in the periods immediately following the rest breaks', *SB* 2 (1927), 4.

Before the time study engineer came round, Pedroni herself used to secretly quietly watch and see how much time we needed, and if we took an hour and he said 'No they took forty minutes', she used to argue about it, 'They took an hour and you've got to let them have an hour'. As far as that was concerned you really had to respect that woman. (Maria V.)

Not all the women liked working in the stifling world of the Winding Shop with its innumerable rules and regulations and strict discipline, but for certain of them the security of an all-female environment, relatively good pay and conditions, a pride in their skill at a job that was still mostly manual, and a feeling that even though Pedroni was very authoritarian at least she was just, meant that they were pleased and proud to be winders. Some of the winders did feel, therefore, that their interests were cared for and, although it was a disciplined world, the rules were clear. As will be discussed in Chapter 9, their accounts of their lives outside the factory show that they were used to very little freedom in the way that they behaved, and the all-female environment protected their 'respectability'. For many of them the rules and discipline seemed an acceptable price to pay for the benefits of being a winder and thereby a member of a female élite within the factory.

'ONLY THOSE WHO DO THEIR DUTY HAVE RIGHTS'. OPINIONS AND ATTITUDES TO WORK AND FACTORY

Not only the winders but almost all of the female interviewees from the various different shops of the factory remembered their work with pleasure. This was true also of some of the men, but others such as Santino, a skilled worker who started work at Magneti in 1935, expressed a different opinion. Santino, who had been previously employed in the German firm OLAP in Milan, described the conditions of work as far more alienating at Magneti:

there the worker felt more like a slave, because he was treated like a cog in a wheel. Those gentlemen in white shirts used to come along with a chronometer in their hands and tac! they used to start the watch . . . the worker didn't feel like a man any more but like a part of an engine that was working to a set pattern.

But amongst the female interviewees this stress on the inhuman aspects of factory and work was strikingly absent. This was probably because, unlike some of the men who had previous experience of engineering firms with a 'more traditional' form of work organization, for most of the women this was a new type of work and women could not look longingly back on a past when they had been skilled workers with far more control over the work process.

Although it is clear to some extent that such recollections were influenced by the fact that many of the interviewees were far from happy with their current situation as pensioners on a low income, many of them living alone, these positive memories seem to be the result of more than just nostalgia. As Bruna commented, 'you get tired of being at home all the time'. Delfina remembered:

> Oh, I loved it. To tell the truth I really loved it. When I had to stay at home I was really sad. When I was married and had my daughter and had to stay at home and do housework and cook I used to hear the factory bell ringing at five o'clock and I always felt really unhappy.

These women did, therefore, see work outside the home as a broadening and generally positive experience. But it was not just the fact of working outside the home and contact with the outside world which was seen as good; more specifically most of the women considered that Magneti itself was a particularly good place to work. Women's work in this firm was mostly clean and not dangerous, except in the notorious Shearing Shop. Those health and safety problems that did exist were on the whole taken seriously by the management (or at least seen to be taken seriously) and there was a wide range of recreational and welfare facilities. As Delfina commented, 'there was the canteen, the nursery. We weren't treated badly.' The work was almost all done sitting down and was mostly not physically demanding.

Quite apart from the welfare provisions, work in the light engineering sector was, undoubtedly, highly desirable compared to most of the work open to women in this period. As I have argued above, women in the Sesto area had much more restricted job opportunities in industry compared with men. Maria G. felt that 'in those days engineering offered some of the best jobs open to women'. In the countryside near Sesto there was work in the textile industry, which employed huge numbers of female workers, but there the work was heavy, the hours long, and the pay low.[11] Maria V., for example, had worked in a silk spinning mill before Magneti. She said 'a mill is a terrible place to work, because you have to work in hot water, boiling in fact, because if the water isn't boiling then the thread won't detach itself from the cocoon.' She felt that the comparison between the two jobs is one reason she liked Magneti so much:

> I was used to that dreadful job in the mill, so when I started work at Magneti it seemed like gold to me, a really beautiful place. While I was learning I was really afraid that they might send me home so I put everything into the work. I put a hundred per cent of my energy into it. I was really motivated to learn.

There were also job opportunities for women outside the industrial sector, but they too were, on the whole, less attractive than employment in light engineering. Domestic service was one option, but this was, of course, seen as inferior, at times no more than a kind of stop-gap job whilst waiting for something better to turn up. Luciana, for example, mentioned that when she lost her job in the box-making factory Scatolificio Ambrogiano, 'While I was waiting I did a few hours domestic service, because I couldn't hang around doing nothing.' Another possible job for women was dressmaking but, although undoubtedly a more desirable type of occupation than domestic service, it meant lower pay than at Magneti and a long unpaid apprenticeship was necessary.[12]

Indeed the desire to start earning immediately was cited by a number of the women as their reason for looking for factory work. For some the motive was purely economic:

> my father didn't want me to do factory work, he wanted me to go in for dressmaking and in fact I really was quite good at sewing and embroidery. But as there wasn't much money . . . My brother was already working at Magneti and, as he knew the head of the Personnel Office, they gave me a job. (Delfina)

[11] On conditions in the textile industry in this period see B. Bianchi, 'I tessili: lavoro, salute, conflitti', in *Annali Feltrinelli* (1979–80), 973–1070.

[12] On dressmakers see V. Maher, 'Un mestiere da raccontare: sarte e sartine fra le due guerre', in *Memoria*, 8 (1983).

Others present their decision to seek work at Magneti as more of a deliberate personal choice:

I ended up as a worker but if I'd wanted I could have stayed on at school. As I said before, I started work as a dressmaker but I didn't like it, so I ended up at Magneti because you started earning straight away there. (Carla)

Earning wages conferred this kind of prestige, a feeling of self-importance. Pierina sums this up well. Her parents did not send her to the factory but she herself chose to start work, proud to earn and contribute to the family budget:

I wanted to earn my keep to be . . . in those days the word independent didn't exist, but I wanted to be independent, earn my keep and bring a wage packet home. I remember holding on to my first wage packet tight as anything, because I was afraid of losing it. I was really proud to be able to help my family.

Not only Pierina, but the majority of the female interviewees presented work and the ability to earn as fundamental to human dignity and self-esteem. The attitude that all of these women had to work is noteworthy:

I really wanted to work. First to help my family. I would have stayed at home if I'd had work to do there like dressmaking but, stay at home and do nothing! No! I'd have felt like a thief if I'd stayed at home doing nothing. I never liked doing nothing. (Giuseppina)

Like male workers in this period, these women considered work to be a sacred duty and such an important part of life that to question it was unthinkable. As Pierina said:

We had to work if we wanted to eat, we really felt as if we had to work. Life was like that those days, anyway, there weren't any entertainments, nothing much to do. Work was our aim in life, for us it was everything.

We had to work. Its true they didn't pay us very much but in my opinion it was right to say 'You come here to work and if you want to be paid you have to work hard'. (Delfina)[13]

Perhaps the most vivid summing up of this attitude was the comment by Maria G. that

[13] This was the only reference that any of the women made to the pay being low. I think that Delfina was comparing the pay with today's wages rather than specifically with other firms of the time.

Pedroni was very strict, a very strict woman. She expected you to simply get on with your work without making a fuss and wasting time. She used to say 'I make sure you earn well but in return you have to work hard.' And in my opinion she was right. I remember that there was a huge notice over the door saying 'Only those who do their duty have rights', and its really true . . . That was something that Mussolini said, 'Only those who do their duty have rights', and it's really important because you can't expect something for nothing.

Intrinsic to this deeply held work ethic was a pride in their ability and resistance:

They never moved me to another job. They used to sack workers who didn't work hard enough. But, as I always did my duty, they never wanted to send me away. (Olga)

I was quick and even when they put me on a new job thinking that I wouldn't be able to do it, I showed them all up, even the foreman. (Gina)

Yet, despite this insistence on their capacity as good workers, these women generally lacked ambition. As the previous chapter showed, women did not threaten the male career structure by aspiring to take over male jobs. But even within their own work sphere they do not seem to have welcomed promotion to those opportunities for advancement that were open to them. In contrast to many of the men, who hoped to rise up in the factory, most women seem to have seen promotion as more of a burden than a blessing. Orsola, overlooker in the Assembly Shop, initially tried to refuse promotion:

I said to the foreman 'I don't want to do it, put me on the assembly line with the others and I'll do the same work as them.' 'No, no, no!', he said, 'I'll take you to the director.' I replied that he could take me there if he wanted, 'I'm not refusing to work, I'm just refusing to do this job because I've got a home to take care of.'

Delfina, who worked as a quality control inspector during the first period she was at Magneti, returned to work after having her first child and

they put me straight on the assembly line. I wanted to work on the line because there were too many problems, too much responsibility [in quality control]. When the products didn't work properly you were told off by the head of shop and so I said that I'd rather go onto the line as it

: work itself I liked, putting the condenser in position, welding, ne wires . . .

The women did not see work, therefore, as a 'career' in any sense, and once they had learned their job they disliked change or promotion.[14] Undoubtedly this was partly a result of the rather unappealing nature of the type of promotion open to women. Becoming an overlooker or quality control inspector was very different from the potential heights to which men could aspire. Equally importantly, the rejection of promotion must be seen in the context of women's dual burden of domestic and factory work. With household responsibilities women were reluctant to take on additional problems at work. Women's reticence to seek advancement was also in all probability a result of the different expectations that women and men had been brought up with. Women did not generally grow up with the idea that they were supposed to have a career, and Fascist propaganda, which tolerated working-class women's employment only in as far as it could be seen as an extension of their duty to the family, only reinforced this idea.

THE 'GOLDEN FACTORY'

It might be expected that women would recount their experience of working in such an apparently dehumanized factory environment as this in a negative fashion. Their work was generally monotonous and repetitive, and mostly required little skill. Discipline was rigid throughout the factory and especially so in the all-female Winding Shop. The problems with the pay system described in Chapter 4 weighed as much upon the women as the men, but even the best-paid woman earned less than the worst-paid man. In addition, there were frequent problems of jealousy engendered by a combination of the organization of the

[14] The interviews even suggest that women preferred to remain in the shop where they started work initially, amongst workmates that they knew. This resistance to change was also noted in an article in the factory newspaper in 1927, since the management saw it as a problem for the 'scientific organization of the work process': 'we have noticed that many of the workers, and particularly the women, seem to be hostile to any changes in their work.' See 'Le possibilità pratiche di organizzazione scientifica del lavoro in Italia. (Il fattore umano)', *SB* 5 (1927), 23.

work process, which fostered mutual suspicion and rivalry, and women's own insecurity in the workplace.

In spite of all this, most of the women I interviewed had good and positive memories of working at Magneti in the inter-war period and one of them, Maria G., even went so far as to call Magneti 'a golden factory'. Undoubtedly nostalgia and their frequently unhappy current situation as pensioners contributed to this, but the full explanation is more complex. A number of factors must be taken into consideration, such as the nature of the work itself (clean and not very demanding physically, except for winding, which was skilled enough to generate a sense of pride), the very favourable comparison with other types of work available to women with no educational qualifications, a narrowness of experience of life beyond the confines of the home, and, most importantly, a strong belief in the value and dignity of work.

The attitudes and opinions of the women I interviewed cannot be assumed to be representative of the work-force as a whole, but they do indicate that at least some of the female workers had internalized the values of discipline of the factory and adapted to the conditions of work, accepting their situation and sometimes even praising it. Here, it is likely, lies at least part of the explanation for the difference in turnover of men and women described in an earlier chapter.

But the loyalty and disciplined attitude of these women should not be seen as purely emanating from their particular situation as women. In addition, it is important not to underestimate the role of management strategy itself in shaping women's (and indeed all the workers') attitude to the firm. Magneti provided a whole series of welfare facilities designed explicitly to win the affection of the work-force and a consensus for the management's organizational strategies. Some were unashamedly of the old traditional paternalist variety, while others were dressed up in more modern 'American' garb. All were presented as being, at least to some degree, for the greater glorification of Mussolini and the Fascist regime.

8

'A Truly Humane and Political Policy': Welfare Provisions

BY the inter-war period many Italian firms, especially larger ones, provided welfare facilities for their work-forces. Perhaps the best-known example in the Milan area was the Crespi textile firm, where workers could benefit from a very full range of provisions including a workers' village in which even the local police station was housed in a firm-owned building.[1] But paternalism was not confined to the textile industry. Even much more modern firms such as Fiat and SIP provided welfare facilities,[2] as did most of the larger firms in Sesto itself. The nearby Falck steelworks had its own workers' housing, the *Villaggio Falck,* as well as numerous other facilities such as a shop, a children's nursery, a workers' hostel, and a canteen.[3] In the Sesto plant of the Breda heavy engineering firm there were similar facilities including workers' housing, a canteen, a hostel for male employees, and a shop.[4] Crespi, Falck, and Breda were perhaps particularly good examples, but they were far from alone. According to an official survey, by 1927 in the province of Milan forty firms provided workers' housing, ten had dormitories,

[1] 'Le opere di assistenza sociale create dagli industriali. L'industriale Cristoforo Benigno Crespi. Fonda nel 1878 in una landa deserta la borgata industriale di Crespi d'Adda', *L'Assistenza sociale nell'industria* (ASNI), 3 (1927), 5–9. On the paternalist initiatives of certain textile firms, including Crespi, see L. Guiotto, *La fabbrica totale: paternalismo industriale e città sociale in Italia* (Milan, 1979). See also G. Baglioni, 'La costruzione del paternalismo organico nel pensiero di un imprenditore d'eccezione: A. Rossi', *Studi di sociologia*, 3–4 (1971); Baglioni, *L'ideologia della borghesia industriale nell'Italia liberale* (Turin, 1974); S. Merli, *Proletariato di fabbrica e capitalismo industriale* (Florence, 1972), i. 357–72.

[2] V. De Grazia, *The Culture of Consent: Mass Organization of Leisure in Fascist Italy* (Cambridge, 1981). On Fiat see also 'Le vaste iniziative della Fiat', ASNI, 3 (1928), 5–16.

[3] On the Falck social welfare provisions, see 'Acciaierie e Ferriere Lombarde Falck', in Fascio di combattimento di Sesto San Giovanni, *Sesto San Giovanni fascista industriale* (Milan, 1934), 34–45.

[4] 'Breda', ASNI, 1 (1928), 8–13.

fifteen had their own sports fields, thirteen had libraries, fifteen had nurseries, and four ran domestic science schools.[5] The welfare facilities of Magneti Marelli were not, therefore, particularly unusual in this period.

At Magneti, apart from the more clearly productivist welfare initiatives described in earlier chapters, there were also numerous other facilities such as a nursery and a canteen, which were more like the classic paternalism practised in other less Americanized firms. Ercole Marelli himself had in fact been an old-style paternalist, and this tradition was carried on in the new firm.

The Magneti management's keen interest in welfare issues was undoubtedly linked to its desire to encourage a low turnover of the work-force, but it seems clear that it was also due in part to the large number of women employed there. Welfare facilities were often particularly good in firms with a female labour force. Some, for example, provided hostels for young female employees so they could live away from home in a 'respectable' environment next to the factory.[6] Initiatives like these were generally extremely value-loaded, with the aim of reinforcing norms of femininity and domesticity. Their role was to ensure that women's 'morality' and childbearing capacities would not be impaired by factory work. More specifically, in the political climate of the inter-war period, provisions such as factory nurseries helped deflect potential criticism of firms which made extensive use of female labour.

Welfare policies were also important in making the Fascist mirage of 'class co-operation' appear to work by making it look as if the working class gained concretely from Fascist labour

[5] G. A. Vigliani, 'L'assistenza sociale nell'industria italiana', ASNI, 1 (1928), 2–6. The figures in this survey show that Milanese industry generally had good provisions compared with most of the other Italian provinces. To give one example of this, there were only 46 nurseries run by firms in 1927, which means that a third of all of these were in the province of Milan.

[6] This was unnecessary at Magneti due to the availability of local female labour who could all live at home. In those firms where this type of hostel did exist they were usually religiously and moralistically run, with a strong emphasis on domesticity. See e.g. for a description of the hostel of the Snia Viscosa textile factory in Rome, V. Delmati, 'La società generale italiana viscosa—Roma', ANSI, 2 (1929), 23–31. On similar hostels in the Milanese area see A. Ariano *et al.*, 'I dormitori-convitti e i villaggi operai in una zona tessile del nord Milano', *Classe*, 14 (1977), 145–64.

policies. Welfare provisions tightened a firm's grip on its work-force and at the same time provided a good and generous image of management. This was recognized even by so-called 'left wing' Fascists who opposed the employers' enthusiasm for welfare policies. The employers' intentions were in any case explicit, as is evident in the following extract from a speech given in 1931 by Benni, the Magneti president:

> It is pointless simply hoping that the workers can automatically understand the concept of class cooperation if they cannot see any concrete results . . . For the past few decades workers have hated their employers and this will not simply disappear to make way for cooperation unless outside the shopfloor (where the hierarchy must be respected to maintain discipline) the employer pays attention to his workers' spiritual and physical needs, and attempts to alleviate their troubles, if in short he does not try to make his firm just like one big family. But this does not simply mean 'paternalism', as some call factory welfare policies. No, this is a truly humane and political policy which helps to build up our workers' understanding of and faith in the Corporate State.[7]

According to certain articles in the factory newspaper, the Magneti welfare policy differed from classic paternalism also in another respect. As Chapter 3 has shown, the Magneti approach to rationalization contained a strong emphasis on 'human engineering', and during the period of greatest apparent enthusiasm for scientific management, the 1920s, the management tried to present even these less obviously productivist initiatives as being also, in some ways, part and parcel of the 'Americanization' of production:

> In addition we try to win our workers' affection and make them take an active part in the firm through an appropriate welfare policy . . . By welfare we don't simply mean useless philanthropy or charity but a means of actually improving output and also a kind of compensation for the amount of work expected of them.[8]

In spite of such claims, in contrast to policies like the industrial health campaign, in practice there was not a great deal

[7] A. Benni, 'L'assistenza ai lavoratori. (Discorso dell'On Benni sul Bilancio delle Corporazioni seduta del 6 maggio 1931)', ASNI, 3 (1931), 43.

[8] U. Quintavalle, 'Le possibilità pratiche di organizzazione scientifica del lavoro in Italia', *SB* 5 (1927), 24.

of difference between the kind of welfare provisions described in this chapter and those in more traditionally paternalist factories such as in the textile industry. The initiatives considered here did, however, share a one common aim with the more 'scientific' ones discussed in Chapter 3, that of creating a consensus amongst the work-force for the firm and its Americanist innovations. As the firm's newspaper argued in 1925:

> All the welfare, humanitarian, and recreational facilities that we have created and are enlarging daily, clearly cost us a great deal of money: but we accept these sacrifices gladly since, in addition to the legitimate moral satisfaction that we feel in caring for the workers who have helped make this firm great, we are also compensated (or we shall be compensated) by their greater loyalty to the firm and greater output.
>
> The strength of an industry like ours lies, to a great extent, in its work-force: motivating the workers means encouraging them to increase their productive capacity, which is also to their own advantage: it implies establishing a more caring relationship between firm and work-force so that they can both count on each other in times of need.[9]

THE FACTORY SOCIAL WORKER

From the early 1920s Magneti Marelli employed its own social worker. The Fascist state actively promoted the spread of social workers in industry. Social work was considered a suitable profession for middle-class women, being seen as an extension of women's 'natural' nurturing and caring role. It was also more acceptable than other types of work where women might 'rob' men of their jobs.[10] To encourage factories to employ social workers, a special school was founded in 1927 to train them, the *Scuola superiore femminile fascista di assistenza sociale*. This accepted women between the ages of 21 and 45 who were 'physically healthy, of good moral conduct and spinsters or childless

[9] *SB* 7 (1925), 11.

[10] As an article in ASNI argued in 1928, the expansion of the number of women employed in this way 'could also help resolve the problem of women's employment, especially with regard to those still far too numerous women who are trying to get into the liberal professions, which have been up till now an almost exclusively male domain' (P. T., 'Il servizio sociale nelle industrie', ASNI, 3 (1928), 3).

widows'.[11] They had to have either secondary education or at least a nurse's diploma. Some of the actual trainees even had university degrees. Undoubtedly all were of middle-class origin. The eight-month diploma offered by the school included subjects like hygiene, nursery care, domestic science, and courses about the effects of scientific management on workers' health. In addition, each future social worker had to study the 'Corporate state' and other aspects of 'Fascist theory'.

The political ideas of these women were clearly important, since they were in an ideal position to convey Fascist ideology to the masses. Their actual work in a firm was often only vaguely defined, so it was essential to ensure that they were good Fascists to avoid embarrassing situations where they might become champions of the work-force. Trained in this way, these women did not shy from politics. As Lia Tarugi, a one-time student of this school and later Inspector of Factory Welfare Services, wrote, the job of the Fascist social worker was to encourage the workers to see the management's point of view. She was to listen to their complaints and then explain to them why they were wrong. She should

> convince them that there is a reason behind everything the firm does, show them how tomorrow's well-being depends on today's sacrifice, explain to them the role of the State . . . get them to understand and value everything that the Fascist Government has done, is doing, and will do for the health and well-being of the working masses.[12]

At Magneti this post was filled throughout the entire Fascist period by a certain Evelina Garbagnati. Like other factory social workers, she appears to have fulfilled her political role and at the fall of Fascism was among those 'sent away' from the firm.[13] Arguably she had more power than any other woman in the firm (with the possible exception of Luigia Pedroni). Her job included organizing and co-ordinating all the factory's welfare initiatives apart from the more clearly productivist ones, such as

[11] 'La scuola superiore femminile fascista di assistenza sociale', in ASNI, 5 (1928), 8.

[12] L. Tarugi, 'Assistenza sociale fascista di fabbrica', in ASNI, 4 (1930), 91.

[13] Garbagnati was not amongst those sacked 'for Fascist activities'. The reason for her dismissal was described rather as 'because considered to be undesirable and therefore harmful to the good functioning of the firm' (Comunicazione 07, CLN aziendale FIMM Stab. A, 9 May 1945, ACLNL, b. 4, f. 13).

the health and safety policy, which were the domain of the firm's doctor and medical centre. She was, in many ways, the linchpin of the firm's consensus strategy, providing a safety valve for complaints and offering individual rather than collective help in times of need. Her influence spread throughout the factory, even into the workers' homes, which she regularly visited:

Since, unfortunately, we are not able to provide actual workers' housing we try to keep ourselves as well informed as possible about the living conditions of our work-force and their families through regular home visits by the Social Worker. This has proved an excellent method of discovering our employees needs and above all *of penetrating our workers' psychology*.[14] (my italics)

One of the women I interviewed described her role in the following manner:

[Garbagnati] wasn't really a social worker. She did do a few things but she wasn't . . . She used to help people who were in need . . . it was like nowadays . . . someone might go to a friend for help, for example, and that friend might herself go to the social worker on your behalf. That was the sort of thing Garbagnati did . . . She wasn't a real social worker, no . . . Because those people weren't social workers, they were something else, they were like the San Vincenzo charity workers. Garbagnati and her lot were sort of the bosses' helpers. You know they used to look after the interests of the workers and the bosses. They helped people who were really in need. A real social worker does a totally different job from that kind of individual help. (Orsola)

Orsola's comments accurately pinpointed the essential characteristics of the social worker's job. She appeared as a charitable lady lending a benevolent ear to individual problems. Her work was to filter complaints and speak on behalf of the workers even though she was, as Orsola pointed out, the 'management's helper'.

Much of Orsola's understanding of Garbagnati's role was undoubtedly due to hindsight. She described the limitations of the social worker only in terms of the different role of social workers since the war. It is not easy to tell what the attitude of the workers was at the time, although some of the interviewees did argue that her sacking in 1945 was wrong:

[14] U. Quintavalle, 'Le possibilità pratiche', 24.

[Q. Were they well looked after in the nursery?] Yes, very well. There was a lady who did a great deal of good in my opinion. They sent her away during the Insurrection and I've always said that it was a mistake to get rid of her. She wasn't a bad person. She was the organizer. Signorina Garbagnati she was called, and she was so good at her job . . . (Olga)

THE WORK OF THE WELFARE OFFICE

Garbagnati was the head of the firm's Welfare Office. This ran a wide range of provisions. There was, for example, a workers' canteen with separate facilities for men and women, and a more luxurious one where the clerical staff could dine off white tablecloths. The workers' canteen does not seem to have been particularly cheap. In 1927 the factory newspaper quotes the cost of a full meal there as 3,70 lire,[15] which was expensive when compared with the 8–9 lire per day that many of the women earned in this period. But this price was for a fairly large meal consisting of a soup course, meat and vegetables, bread, cheese or fruit, and coffee. Very few workers ate anywhere near this quantity at home. Oral testimony revealed that many families still had extremely poor diets in this period and this might have been food for at least a day, not just one meal. Because of the expense it is highly probable that a situation arose similar to that in the Ercole Marelli factory, where it was usually only the men who ate in the canteen, women instead bringing their own lunches of poor remnants of the previous evening's meal.[16] Those who did eat in the canteen probably usually only ate a single course.

For those too poor for the canteen, the firm provided other ways of keeping the cost of living down. One such initiative was the provision of allotments for workers so that they could

[15] 'La mensa e la sala convegno', *SB* 2 (1927), 24.

[16] These women's decision to do without a hot meal as they desperately struggled to make ends meet was praised in the following nauseating manner in the Fascist press: 'The women who are more thrifty than the men mostly had their lunches with them, usually left-overs from their modest evening meals. The Woman of the poorer classes is always ready to sacrifice herself. How much delicacy and poetry there is in this instinctive self-denial' (P. Preda, 'Una visita alle opere assistenziali della Società Ercole Marelli', ASNI, 5 (1932), 84).

supplement their earnings by growing extra food.[17] Similar to this was the firm's shop, which sold food at low prices, thereby helping the firm to justify wage cuts. Indeed the Fascist government and the *Confindustria* (of which, of course, Benni was the president) encouraged the provision of factory shops in the late 1920s for precisely this reason.[18] These two initiatives highlight how welfare policies could be seen as a replacement for Taylor's high-wage strategy.

Extra nourishment was also provided for certain workers, who were given supplementary meals of milk and egg designed to help guard against TB. Sometimes the workers were chosen for this because they seemed particularly sickly, but in the late 1920s the entire work-force of the Winding Shop was automatically getting an extra meal of milk, possibly as a protection against the bakelite fumes found there.

Various other initiatives were aimed specifically at children, either young employees or the daughters and sons of workers or clerks. Very young workers were given special rest breaks morning and afternoon, although this may have seemed a dubious pleasure as they received no pay for the period in which they did not work.[19] The children of both male and female workers who had been on the firm's payroll for at least a year could benefit from special holidays at the seaside. These holidays, which lasted a month, were for children aged six to twelve years who had particular health problems. There were also holiday camps organized in the countryside at Monasterolo d'Adda, where the Quintavalle family owned land and had a country residence. This facility was provided specifically for the employees of the nearby Magneti subsidiary firm ICS. Although the management were keen on stressing the fact that these initiatives were financed not by the firm but directly from the pockets of the Quintavalle family itself, antifascist sources noted

[17] *SB* 7 (1927), 30.

[18] This is explicit in much writing on the subject in the period of the cost-cutting campaign. See e.g. G. Gobbi, 'Assestamento industriale e assistenza sociale', ASNI, 1 (1931), 3. The Fascist unions in Milan felt particularly threatened by this part of the employers' strategy, partly because of this cost-cutting aspect and also because the unions found themselves specifically forbidden to set up their own shops (G. Consonni and G. Tonon, 'Milano, classe e metropoli tra due economie di guerra', *Annali Feltrinelli* (1979–80), 468).

[19] 'Dalla Magneti Marelli di Sesto San Giovanni', *Unità* (31 Mar. 1926), 3.

that in fact half the cost of the holiday was deducted from the parents' pay packets.[20] To keep costs down volunteer labour was employed as far as possible, and the holiday camp at Monasterolo was run by the social worker, aided only by female members of the Quintavalle family and volunteers recruited from the local *Fasci femminili* (the Fascist women's organization).[21] The holiday camps were run in good Fascist spirit:

> the children are cared for just as if they were part of a large family, lovingly but also with plenty of discipline. Following the GIL regulations, together with education of their minds they have been taught a sense of duty, fraternity, and love of their country.[22] (GIL was a Fascist youth organization.)

> Although the camp regulations are caring and maternal they are also designed to make its members into real *Balilla* [another Fascist youth organization]. These regulations are strictly observed from the first day to the last, from the morning flag-raising ceremony where homage is paid to the Royal Emperor and to the Duce right through until the evening prayer session in which the children softly whisper prayers in praise of the greatness of our Motherland.[23]

Magneti provided a variety of initiatives for the workers and their children. Some were aimed at women alone and others at all of the work-force. Generally the facilities seem to have been comparable to those provided by many other contemporary firms in the area. Where they were exceptionally good was in the field of maternity provisions.

ONMI AND MATERNITY PROVISIONS

The active pro-natalist policy of the Fascist government included the establishment of the Opera Nazionale per la Maternità ed Infanzia, the National Mother and Child Organization (ONMI).[24]

[20] 'Dati e considerazioni sulle condizioni della classe operaia: le condizioni di lavoro degli operai della "Magneti Marelli" ', *Lo Stato Operaio* (15 June 1939).

[21] The links between the local Fascists and the Monasterolo camp were particularly strong, perhaps due to the fact that the managing director of ICS Claudio Politi was also the local *Podestà* (Fascist mayor).

[22] 'La Colonia ricreativa Ida Quintavalle Frulli', *SB* 6 (1938), 44.

[23] 'La Colonia marina Ferdinanda Quintavalle', *SB* 5 (1937), 40.

[24] On ONMI see S. Mongio and M. Rimola, 'Per una definizione del rapporto donne-istituzione: il caso ONMI dalla fondazione al 1935', tesi di laurea (Trento,

Its role was to help mothers care for their children better and to encourage women to devote themselves full time to child care if possible.

The Milanese Federation of ONMI was in some ways untypical, since it was frequently singled out as the province in which this organization functioned at its best. This had some kind of logic in the demographic campaign, as Mussolini had characterized Milan as the most infertile of all the Italian cities.[25] The unusual efficiency of the Milanese ONMI Federation may have been partly due to the good administration of its Special Commissioner Sileno Fabbri,[26] but the real explanation lay in the nature of ONMI itself. ONMI, like many other Fascist organizations, was supposed to get maximum results at minimum cost, even if this meant a certain flexibility and eventual compromise in its aims. This meant reducing direct initiatives to a minimum and concentrating instead on making use of existing provisions. At the stroke of the legislator's pen all the private and public institutions involved in maternity automatically fell under the jurisdiction of the new organization. A relatively wealthy city like Milan was far better provided for than poorer areas in this respect, since richer areas tended to have a greater number of charitable organizations of this nature. ONMI inherited all the unequal distribution of the pre-Fascist maternity facilities, for real redistribution could not be carried out within the severe cash limits which dogged the organization's activities.[27]

Its personnel was largely voluntary, and its mainstay, especially in the local groups (Committees of Patrons), was the army of

1977–8). See also *Maternità ed Infanzia, Bolletino ufficiale dell'Opera Nazionale per la protezione della Maternità e dell'Infanzia* (Rome, 1926–35); G. Festa, *L'Opera Nazionale per la protezione della Maternità e dell'Infanzia dalla sua fondazione* (Rome, 1962).

[25] B. Mussolini, *Discorso dell'ascensione: il regime fascista per la grandezza dell'Italia* (speech given to the Chamber of Deputies, 26 May 1927) (Rome–Milan, 1927), 23.

[26] On the work of Sileno Fabbri as Commissario Straordinario of the Milan Federation from 1928 see numerous documents contained in ASM, Gab. Pref., cart. 378, f. 'ONMI'.

[27] In 1930 Grossi, who was director of the Milanese *Asilo Materno Provinciale* (a special maternity home for unmarried mothers), commented that in spite of the new measures, after 5 years of ONMI activity, 'In general peasant women give birth in the same conditions as animals and the doctor is only called when a forceps is needed or the placenta is not expelled properly' (G. Grossi, *La protezione e la difesa sociale della maternità in regime fascista* (Milan, 1930), 195).

petty bourgeois women who worked as volunteer home visitors. Such work helped fulfil their desire for extradomestic activity whilst firmly retaining them in the acceptable feminine role of nurturer. Much of ONMI's work was to co-ordinate and inspect charities, infusing them with a 'new Fascist spirit' and encouraging them to comply with state regulations by handing out small subsidies. In the philosophy of ONMI, the onus of child care remained firmly with the mother and no attempt was made to relieve her load. Instead ONMI tried to improve her knowledge of how to look after her baby whilst giving her virtually no material assistance, since the aim was to strengthen the family and confirm women in their maternal role. One aspect of this was the attempt to force unmarried mothers to recognize and take responsibility for their babies.

ONMI had an essentially political role and it was a key institution in the transmission of ideology concerning women. If, as has frequently been argued, the Fascist period saw the first attempt in Italy at state intervention in the family,[28] undoubtedly ONMI was one of the most important organizations in this policy. The Fascists brought the question of women's role into the open, focusing public attention on what had hitherto been considered a private matter by many. Paradoxically, women were to be 'mobilized' by this propaganda not to encourage them to take active action in support of Fascism but instead to reduce their horizons and fix their ambitions solely on a career of motherhood and a domestically orientated existence. ONMI's function was to bring this message into every home.

Financial constraints reduced the assistance part of the programme to a minimum, and even the assistance organs functioned mainly as educators and spreaders of propaganda. In spite of ONMI's claim to represent a new and dynamic approach to improving child care, it incorporated all the old class-based humiliating values of charity and the Church's reactionary views on women's place in society.

Compared with most other Italian provinces Milan had quite good maternity facilities. There were four clinics run by the *Cassa di Maternità*, providing outpatient care for pregnant and

[28] C. Saraceno, 'La famiglia operaia sotto il fascismo', *Annali Feltrinelli* (1979–80), 192.

nursing mothers,[29] and ONMI had a number of maternity care centres located in the poorer areas of the city which each contained an obstetric consultancy, a paediatric clinic, and a psychiatric consultancy. These centres also distributed free milk and ran a canteen for pregnant and nursing mothers. The clinics were financed from a mixture of central ONMI funds and private donations. The existence of these clinics, although in their turn hindered by a shortage of funds,[30] is evidence that the provisions in Milan were unusually good. In Sesto itself by 1933 there was a children's clinic run by the *Fascio femminile* and from February 1933 this also functioned as an obstetric and gynaecological clinic for pregnant women for one-and-a-half hours per week. In 1934 the ONMI section also set up a clinic for women who were breast-feeding their babies, financed by the municipal council.[31] In addition there was a private nursery run by nuns.

Relatively speaking, Milanese women were well catered for, but even here ONMI's provisions were limited. Many women were excluded simply by not living near to those facilities that did exist. This was reflected in the testimonies I collected, as some of the women were uncertain whether ONMI existed yet in this period, although a few remember going to the newly built *Policlinico* (hospital). Also there was a drawback with the

[29] These clinics were similar to the ONMI ones but with the difference that they were only open to women who were paid-up members of the *Cassa di Maternità*. This was an insurance fund founded in 1907 to complement the new legislation controlling women's and children's employment, which had made a month's leave compulsory after birth. The new *Cassa* was designed to finance this period, since the employers were not obliged to pay their workers for their time off work. For a modest annual contribution from the worker and her employer, the insured women gained the right to a lump sum on the birth of a child even if it was stillborn. In the Fascist period the *Cassa* was slightly improved, giving e.g. a higher lump sum. It got involved in the demographic campaign and started to award subsidies, ranging from 400 to 1,000 lire, to couples who married before they were 27 years old. See M. Casalini, *La protezione e l'assicurazione della maternità e la difesa della prima infanzia* (Rome, 1934).

[30] 'ONMI has had to restrict the activities of its own clinics so that they limit themselves to re-education and trying to improve hygienic standards rather than actually practising therapeutic medicine . . . ONMI's obstetric clinics, therefore, are seen as complementary to the local hospital facilities and their main role is to watch over the hygiene conditions of pregnant women, and the children's clinics mostly just supervise children's diets and try to teach the concept of hygiene to the mothers' (Grossi, *La protezione*, 123).

[31] Fascio di combattimento di S.S.G., *Sesto San Giovanni*, 17.

ONMI provisions in that very few of their clinics contained nurseries. This meant that a workplace nursery was crucial.

A few Sesto firms such as Ercole Marelli, Falck, and Magneti set up their own day nurseries, but in industry as a whole maternity provisions varied greatly. In the mid-1920s a survey by ONMI revealed a dismal picture. It showed not only that virtually every Italian firm had ignored their legal obligation to provide breast-feeding facilities, but also that women often lost their jobs as soon as they got pregnant. This was stressed in a letter written to *Unità* in 1926 by a local doctor, who pointed out that of approximately 3,000 firms in the province of Milan which employed over fifty women (and were therefore legally obliged to provide facilities), only three actually had these facilities and of these two used the rooms simultaneously as warehouses.[32] By 1930 the situation had scarcely improved and the ONMI inspectors reported that

> even today there are firms in which there are no nursing rooms, or, where these do exist, they are so badly maintained that the new mothers do not want to make use of them . . . There is just as much neglect of the needs of pregnant workers in the last few days before the actual birth. Even though every single pregnant woman ought to be examined medically in order to decide whether or not she is suitable for work in the period just before the birth, this is often not done properly and there is no supervision to watch out for problems that might arise as a result of the pregnancy.[33]

THE MAGNETI MATERNITY FACILITIES

As was shown in Chapter 5, about a quarter of the female workers at Magneti were already married when they joined the firm. Many others probably subsequently married during their period of employment. Some became mothers and a number of these stayed on at work after at least the birth of one child. It seems very likely that some women managed to plan the size of their families, at least in the sense of reducing the number of children, for, despite the fact that contraception and abortion

[32] 'La protezione della maternità nella grande Milano', *Unità* (9 Oct. 1926), 3.
[33] Grossi, *La protezione*, 147.

were illegal,[34] many married women in the urban areas of the north did not have many children. This suggests that the law was frequently broken and birth control quite widely practised. Abortion seems to have been fairly widespread, as can be seen from the continuing frequency of trials of illegal abortionists and their clients throughout the Fascist period.[35] Regarding Magneti Marelli, however, it has not proved possible to obtain any information on whether birth-control methods were widely used by the female workers. In the interviews only the briefest veiled type of reference ever appeared to such matters, for even today speaking frankly about sexuality seems to be seen as taboo by this generation of women.[36] Carla, for example, said

> I had my daughter straight away even though we would have been better without a child at that point, but when you are young! . . . I only had that one child . . . we tried to avoid having her, but these are mysteries, aren't they?

[34] On abortion legislation see D. Detriagache, 'L'Italie fasciste e la répression de l'avortement', *Mélanges de l'École Française de Rome* (1980).

[35] Illegal abortionists were being constantly discovered in Milan. These cases recorded by the police were clearly only the tip of the iceberg, as even most of those arrested had been carrying out abortions undiscovered for some time before coming to the attention of the police (ASM, Fondo Gab. di Pref., cart. 173, f. 'Aborti e infanticidi'). In 1941 135 abortion cases were tried by the judicial authorities in Milan out of a total of 892 for the whole of Italy. It appears that the Milanese judiciary was slightly more lenient than in many other areas of Italy, which suggests that local public opinion was less hostile to abortion than the legislators were (see Detriagache, 'L'Italie fasciste', 733–5). See also the work of Luisa Passerini ('Donne operaie e aborto nel Torino fascista', *Italia contemporanea*, 151–2 (1983)), who demonstrates that abortion was indeed widely used as a means of controlling fertility amongst urban working-class women in Turin.

[36] It is important to note, however, that I decided not to press this point in the interviews, preferring to respect these women's privacy, as it was obviously a sensitive subject and would have seemed arrogant and perhaps insulting on my part to have insisted. Possibly an older woman doing interviews of this kind might feel more at ease asking this type of question, as she might feel closer in life experience to these women. Because of my relative youth and my unmarried status these women may have been even more than usually reticent to discuss such questions with me, perhaps fearing that they might shock me. I wanted the interviews to be more of a sharing of information than an interrogation, and respect seemed more important than getting every last drop of information. The very language used by these women to describe anything sexual does, furthermore, suggest that this was not openly talked about. In this language babies are not born but *comperati* (bought), and the onset of contractions is euphemistically referred to as *ammalarsi* (falling ill). Rather than saying that they had slept with a man the term used was *vedere* (to see). One woman even said that when she had her baby, her 19-year-old younger sister was actually sent away from the house lest she learn some of the mysteries of sex.

When asked whether the women in the factory ever talked about contraception on the shop floor Olga said

No, we didn't talk, we used to talk about it a little after the war and we used to tell each other a few stupid things. As for me, to tell you the truth, I was so ignorant about certain things, because sometimes they used to tell jokes and laugh and say: 'Olga, come on, don't you really understand?' But there really were so many things that I didn't understand, they used to laugh at me.

Whether birth control was used or not, what is clear is that at least some of the women workers in this firm were unable or unwilling to prevent pregnancy whilst employed at the firm. Far from all of them could count on help from female relatives to assist them in combining motherhood with factory work. For many young mothers in the firm it would have been impossible to continue working without the factory maternity provisions.

In contrast to the vast majority of Italian women, and even to most women in Sesto San Giovanni, the Magneti workers enjoyed good provisions. In the early 1920s a room was reserved to enable women to breast-feed their children during work hours. In 1927 a nursing mothers' clinic was introduced for the use of all women workers with babies. This gave weekly medical checks for child and mother and distributed food and clothing. Garbagnati kept a record of their progress and

the Social Worker visits the women at home to check that all her advice is being followed. The mothers are also each given a little 'catechism' containing all the information they need which the Clinic itself has produced.[37]

Antenatal care and education were provided and also, according to the management, pregnant women were prevented from working in 'unhealthy' parts of the factory.[38]

In 1933 a nursery was opened with a ceremony attended by a delegation of directors, Paolo Quintavalle Marelli, and a priest who gave a blessing. The factory newspaper heralded the event:

[37] 'Il consultorio lattanti', *SB* 10 (1927), 27. Between 1927 and 1933, according to the firm, the *consultorio lattanti* helped 160 nursing mothers, there were 22,050 medical checks carried out, 250 tins of powdered milk distributed, and 1,880 sets of clothing. 15 prizes were also given out to the 'best nursing mothers' ('L'inaugurazione del "Nido" Magneti Marelli', *SB* 3 (1933), 54.)

[38] *SB* 10 (1927), 27.

> This nursery represents—over and above all the other benefits—absolute peace of mind for worker mothers as far as their children are concerned. No longer do they have to be abandoned alone at home or in the inadequate care of their brothers or sisters, or even worse, in the mercenary care of a childminder. Instead, there are lovingly watched over and cared for in perfectly clean, hygienic surroundings.[39]

This nursery provided a free day care service for the children of female employees (the children and wives of male employees could attend the clinic but not make use of the nursery). Children could be placed in the nursery when only a few weeks old. Mothers took them there in the mornings, and they were changed into clean clothes and looked after all day. At certain fixed times of the day, mothers were allowed to come and breast-feed their babies.

These provisions made those of the Magneti women who had children amongst the most privileged working-class mothers in Italy. Care and education was available to them right from the ante-natal period until the child was a few years old. All the women I interviewed spoke well of the maternity provisions, recognizing how well treated they were in this respect. They remembered that the laws relating to maternity were respected and women were not sacked simply because they had a child, as happened in many other firms in this period. Around the time of the birth they were able to take the full forty days stipulated by the law. One woman, for example, remembered that when she was pregnant

> we used to go to Sesto, they even paid our tram fares, they gave us something to eat up until the ninth month—because it was during the war and there was nothing to eat. Magneti was a wonderful factory in those days. (Maria G.)

The good facilities at Magneti had the effect of ensuring loyal behaviour of at least a section of the work-force by making them afraid to lose their jobs; such good facilities were hard to find elsewhere. The nursery represented, far more than any other welfare initiative, a powerful means to co-opt the work-force. Other provisions such as the leisure facilities may have been appealing to some of the work-force, but a nursery was literally essential to certain women. One of the interviewees remembered

[39] 'L'inaugurazione del "Nido" ', 54.

that when women were asked to give up their marriage rings for the war effort the first women to offer them were those with babies in the nursery:

I remember when there was the wedding ring business . . . [Luigia Pedroni] scared us. She sent someone to tell us that if we didn't give up our rings then we'd be sacked and even worse. And my husband said to me: 'It's better if you get the sack or something . . . we'll work something out together . . .' But on the actual day when we were supposed to go down and actually take the rings she didn't force us . . . She didn't make us or say 'You must go down', and in fact there were ten of them who went down, the ten from our shop who had children in the nursery. (Carla)

Furthermore, the opportunity of using even the nursery for (rather crude) political indoctrination was not missed. In 1936 all the children were given 'pre-registration' membership cards to the Fascist youth organizations as 'Pre-Balilla' and 'Pre-Piccole Italiane': 'The giving out of the *Pre-Balilla* and *Pre-Piccole Italiane* membership cards, was the occasion for a glowingly "patriotic demonstration".'[40] The mothers were not consulted, but instead

the Director of the nursery *explained* to the mothers of the new members and symbolic importance of the cards which their children had just been given and the moral and material advantages they could mean.

Amidst prolonged fervent applause to the King and the Duce, the Patroness of the Nursery Donna Paola Quintavalle gave out the cards to the dear little saplings of the nursery, who were taken out of their innocent cots adorned with tricolour flags and garlands of flowers and held in the arms of their proud and happy mothers.[41] (my italics)

Despite their young age, these toddlers were subjected to indoctrination from the cradle. As one writer in the factory newspaper claimed, 'The older ones are about two years old and give the Roman Fascist salute as they come to greet us.'[42] But having their children dressed as baby Fascists and taught to salute almost as soon as they could walk probably seemed a price worth paying for the best maternity facilities available. In spite of the fact that these were factory workers working outside the home who, in the spirit of the demographic campaign,

[40] 'Una simpatica ceremonia al "Nido" Magneti Marelli', *SB* 4 (1936), 34–5.
[41] Ibid.
[42] 'Il nido d'infanzia', *SB* 3 (1937), 24.

ought to have been encouraged to remain at home and produce children instead of earning wages, they were given a real incentive to go out to work.

Magneti Marelli provided a nursery because it wanted a stable female labour force, but the creation of factory nurseries was also encouraged by ONMI. The provision of nurseries papered over the cracks in the demographic campaign, full of contradictions, enabling women to be both mothers and workers since production required female labour and the demographic campaign needed them as mothers. Fascism could not ban female labour in the private sector as this would have irreparably alienated industrialists, so nurseries represented at least a partial solution.

WINNING THE WORK-FORCE'S CO-OPERATION

The welfare policy of Magneti included a variety of initiatives run by the social worker such as a canteen, holidays for workers' children, and maternity facilities. These provisions were to a great extent comparable to those in many other firms. What was a little more unusual was the extent of the maternity provisions reflecting the needs of the firm for a stable, loyal, female work-force. Most of the welfare facilities were used as an excuse for propaganda and indoctrination for the Fascist state, but the increased political awareness of the masses was only part of the explanation of this policy. More important was the winning of the work-force's co-operation to defuse the tensions which could build up behind the facade of class co-operation and generally to smooth the introduction of Americanism. Some of these measures, like the industrial health policy, were an integral part of the firm's rationalization scheme. Others, despite the Americanist rhetoric that was used to describe them, were more similar to the long-tried paternalist policies in the tradition of Italian textile entrepreneurs like Crespi. In the end both paternalism and 'Americanized welfare' had the same fundamental aim, that of winning the workers' loyalty and affection. To this end, the firm also provided very extensive leisure facilities under the umbrella of the regime's Dopolavoro organization.

9

'Nationalizing the Masses': The Dopolavoro and Women's Leisure

WITH the largest membership figures of the mass organizations, the Opera Nazionale Dopolavoro (OND) represented an important part of the attempt to create a consensus for the Fascist regime.[1] Inspired by the ideas of Mario Giani, an ex-industrialist, the newly-born Fascist trade unions saw the provision of workers' leisure facilities as a means of competing with the socialist organizations which had built up an extensive network of cultural and educational societies, forming a kind of socialist counter-culture. In the early 1920s Fascist violence forced many of these to close and where the facilities and buildings were not physically destroyed they were frequently reopened in the guise of the new Fascist Dopolavoro. In April 1925, aware of the propaganda value of a timely interest in social issues, Mussolini agreed to the Fascist unions' demands for the setting up of a national Dopolavoro organization, the OND, but in doing so he also effectively destroyed the unions' control of it as part of his opportunist strategy to balance out the various political forces which could threaten his personal power base.

The new organization, controlled by Giani, initially had a productivist and largely apolitical image. This won the support of the employers in a general way, although few rushed to actually set up Dopolavoro sections.[2] Giani, however, was soon outmanœuvred in the struggle for influence between the Fascist Party and the unions, and when in April 1927 Augusto Turati, head of the Fascist Party, became OND leader, he sacked Giani and turned the OND into 'a fully fledged auxiliary of the Fascist

[1] The membership figures were 280,000 in 1926, 2,780,000 in 1936, and 5 million in 1940: R. De Felice, *Mussolini il duce*, i: *Gli anni del consenso* (Turin, 1974), 198.

[2] V. De Grazia, *The Culture of Consent: Mass Organization of Leisure in Fascist Italy* (Cambridge, 1981), 36.

Party'.[3] The productivist image, however, remained, since 'the party, mobilized in the "battles" for stabilization of 1926–7, was concerned to present itself as a new political force for the transformation of Italy into a new industrial power'.[4]

But in the 1930s, under the direction of Achille Starace (Turati's successor as OND Special Commissioner from October 1930), it became primarily recreational, concentrating on sports rather than technical instruction. This new image, summed up by Mussolini in the slogan *andare al popolo* (going to the people), reflected the political changes of the early 1930s when the regime was trying to create a mass consensus. The Dopolavoro's relatively depoliticized nature made it attractive to wide sections of the population. In the late 1930s, however, it became once more involved with the regime's campaigns, this time to mobilize support for the Ethiopian war.

The OND, therefore, was shaped by the political changes during the Fascist period. Behind the varying aspirations of its leaders, nevertheless, lay one fundamental aim, that of binding the masses to the new state, breaking down class and local identification to create a national identity. This desire to 'nationalize the masses' was, to some extent, inspired by the belief of many 'first hour Fascists'[5] that the sharpening of class conflict after the First World War was largely due to the failure of the Liberal state to integrate the working classes into the nation. In the case of a factory OND section, however, the situation was more complex since the aims and campaigns of the Fascist Party did not always coincide with the needs and strategies of the individual employer.

SHOP-FLOOR CULTURE AT MAGNETI

Initially, most of Magneti Marelli's recreational facilities were not handed over to the new Dopolavoro section.[6] It is likely that this was because the firm was unwilling to let them be seen simply as a conquest of the Fascist regime, preferring to stress

[3] Ibid. 38.

[4] Ibid. 41.

[5] 'Fascists of the first hour' were those who supported the movement from the early years after the First World War.

[6] This was similar to Fiat's strategy. See De Grazia, *Culture of Consent*, 87–95.

that the generosity of the provisions were particular to this firm. Furthermore, once the Dopolavoro did become active, welfare facilities such as the nursery were kept separate as management initiatives. Only the recreational provisions were left to the Dopolavoro.

The Fascist Dopolavoro only became an active force in the firm from 1934 onwards. There is almost certainly more than one explanation for the decision to expand the facilities in this period. One is the increasing links with the regime as Benni became Minister of Communications and the firm became progressively more dependent on state orders and government policies for its survival and growth. The creation of a flourishing and energetic Dopolavoro section was certainly a good propaganda exercise for the factory management keen to prove their political credentials. Perhaps just as significant was the fact that 1934 saw a visit to Sesto by Mussolini himself.

Where employers wanted to show that they were helping to win over the industrial working class for Fascism, they favoured building up the Dopolavoro rather than the Fascist unions which had proved more intractable. But, defeated in their bid for real influence, the unions too could attempt to make use of the Dopolavoro and it is possible that the expansion was at least in part due to the fact that the firm's union group was increasingly involved in the leisure provisions in the late 1930s. To create some sort of presence in the factory the unions had few alternatives. Involvement in the Dopolavoro, even without overall control, was one way of attempting to win the confidence and support of the workers.

Initially, exclusion from the OND made some union leaders oppose its very existence. At the first interprovincial congress of Fascist unions, for example, Nino Cuzzeri (Fascist metalworkers delegate) argued that the Dopolavoro was being used by the employers to undermine the unions' power on the shop-floor.[7] Later, however, the Fascist unions came to see the Dopolavoro as one area of factory life where they could play an active role. As the next chapter shows, the Magneti unions found it far from easy to operate in this firm, and with the 1934 OND circular which instructed employers to include representatives of the

[7] De Grazia, *Culture of Consent*, 90.

unions in the executive committees of Dopolavoro branches, they seem to have seized upon this opportunity to at least attempt to create a sphere of influence.

Magneti Marelli appears to have complied with this directive; from 1934 onwards the names of representatives of the unions such as Mario Palma and Giovanni Marchesi appeared in the list of committee members annually published in the factory newspaper. The management, however, retained a firm grip on the proceedings[8] and reserved the most important positions for their representatives, mostly Fascist activists such as Michele Soliveri, director of the A Plant, and the engineer Vico d'Incerti.

In 1938, when Alberto Pesenti was nominated to an important post in the Italian African colonies, his successor as Magneti Dopolavoro president was none other than Bruno Antonio Quintavalle himself. This may have been due to the management's desire to tighten control over the activities of the Dopolavoro after the financial crisis of 1937, provoked by the direct intervention of the Provincial OND. Rather than giving an account of this incident the factory newspaper only alludes cryptically to a mysterious crisis, stressing that 'luckily, the events of this period . . . did not last long', and refers the curious reader to the 'minute book of this branch'.[9] The situation was, in fact, very embarrassing for a firm which vaunted the luxurious nature of its welfare and leisure facilities, and the high level of spontaneous OND membership. Those selfsame spontaneous members had complained to Pietro Capoferri, leader of the CNSFI (Fascist industrial workers union) that the firm was deducting ten lire per month from their pay packets as the OND subscription, and maintaining that this extremely high rate had been officially authorized. In his report to the Milanese Fascist Federation, Capoferri

> stressed that even though the firm had tried to deny that membership was compulsory, this was untrue since effectively everyone had to join.

[8] The *consigli direttivi* of factory Dopolavoro sections continued to be nominated by management and ratified by the OND provincial authorities. The control of industrial Dopolavoro sections was only really returned to the unions in 1939, when the nominations began to made by meetings of Dopolavoro members and the power of ratification was given to the unions in each firm.

[9] 'Relazione sulle attività del Dopolavoro Aziendale Magneti Marelli nell'anno 1937—XV', *SB* 1 (1938), 104.

A clerical worker, for example, who did not join the Dopolavoro was looked upon very badly.[10]

He requested that the non-authorized part of the subscription be eliminated, both because of the rising cost of living and because the Dopolavoro 'was in danger of being hated by the workers'.[11] The result of this was that the Provincial OND suspended all the subscriptions on 26 July, making Magneti Marelli one of the very few Dopolavoro sections to be subjected to the direct intervention of the local authorities.

The directors were not, therefore, entirely in control. Unfortunately, it is impossible to tell whether this high rate was decided by the management or due to the ambitious schemes of the union representatives working within too tight a budget. Since the archives of both the firm's Dopolavoro and the Fascist unions have been lost or destroyed, the extent of union influence is unclear.

For whatever reason, 1934 was the year that saw the beginning of a really active OND section. Previous to this some leisure initiatives were provided for the work-force. In the late 1920s, for example, trips to the mountains were arranged by the 'Magneti Marelli Excursion Group'. There was also a sports group, a choir, a workers' common room, a small library, and a hall used as a cinema. Various excursions to beauty spots and places of interest were arranged and there was an annual party for the clerical staff. A few activities, such as a Christmas party for workers' children in 1924 and an outing for four hundred workers in 1928, were run by the *Mutua operai* (Workers' Mutual Society).[12]

[10] ACS, S.P.D., O.N.D., f. 509.016, sf. 'varie', intercettazione telefonica di polizia, 16 Apr. 1937. Cited in De Grazia, *Culture of Consent*, 305 n.

[11] Ibid.

[12] In the 1920s only a few large firms had *Mutue operai* (Workers' Mutual Societies) in Milan. These were financed entirely by contributions from the workers until 1930, when employers began paying half ('Il nuovo ordinamento delle Mutue metallurgiche', *Azione Sindacale*, 20 Sept. 1930). At Magneti, once the workers' movement had been destroyed, the *Mutua* seems to have become fairly inactive. An article in the Communist paper *Unità* written by a Magneti worker in 1926 complained that 'The internal regulations refer to a *Mutua* and money is in fact deducted from each workers' pay-packet, but in reality there is nothing but these deductions. No one knows who the officials of the *Mutua* are, or what its policies are, no one ever sees a financial report. In short, it is a highly mysterious body' (Il corrispondente di fabbrica, 'Dalla Magneti Marelli di Sesto

Meanwhile, in 1926, the Magneti Dopolavoro was, at least in theory, 'founded'. According to an article in *Unità*, the management seems to have been far from sure of a spontaneous rush to join. All the workers were lined up in the courtyard of the factory and then informed that they had joined the Dopolavoro. A week later they were asked to become members formally and in case they were not particularly convinced the firm offered to pay their first year's subscription free. When one of the workers asked what the Dopolavoro was he was simply told that it was an organization that offered cut-price cinema tickets.[13]

After this, two years went by before a grandiose ceremony was organized to officially 'found' the *Gruppo Aziendale* (the group uniting all the Fascist activists in the firm), which in reality had been in existence for some time, and the Dopolavoro. This was basically an excuse for some Fascist pomp and ceremony with symbolic rather than practical significance.[14] After this the firm seems to have considered its duty done for a while, since the leisure facilities did not magically take on the name Dopolavoro overnight and they appear to have remained largely devoid of Fascist ceremony.

In 1934 the firm's Dopolavoro got a new lease of life. Recruitment was stepped up and Magneti Marelli was awarded a special plaque by the Provincial OND as the firm with the highest proportion of OND members in its work-force.[15] In the same year the Uberto Quintavalle Sports Field (named after Bruno Antonio's son) was opened by Mussolini himself and a wide range of activities was organized including football, canoeing, motor-cycling, *tamburello* (a 'traditional' Italian game), tennis, and numerous excursions. By 1936 a special Dopolavoro building had been opened next to the A plant.

'MOBILIZING THE WORKERS FOR FASCISM'

One aspect of the newly active Dopolavoro was the amount of Fascist propaganda in its activities. For example, the annual trip

San Giovanni', *Unità* (31 Mar. 1926), 3). In the late 1920s the Fascist unions seem to have got involved with the *Mutua*, seeing it as means of expanding their influence among the work-force.

[13] 'Il dopolavoro alla Marelli di Sesto', *Unità* (29 Oct. 1926), 3.

[14] 'Una ceremonia solenne', *SB* 4 (1928), 31.

[15] 'L'attività del dopolavoro nell'anno XII', *SB* 1 (1935), 101.

to the country at the time of the cherry harvest was basically a chance for some fresh air, but the organizers arranged for plenty of ceremony to accompany the more mundane activity of walking in the countryside. The Dopolavoro members

> went to place flowers and laurel wreaths on the commemorative monument to Arnaldo Mussolini and on the local war memorial. Moved by the event, they gathered, listening to Cav. Comotti who welcomed the Dopolavoro members with a brief but inspired speech which was so patriotic that it roused the enthusiastic approval of the crowd until they chanted repeatedly *alala* to the Duce and to Fascism.[16]

On another occasion, two years later, when an outing was organized to Cremona,

> the local political authorities were all at the station to welcome the Dopolavoro members . . . As soon as they had arrived, they walked in file up to the 'Palace of the Fascist Revolution' preceded by the Works' Musical Band and by two men carrying a large laurel wreath.

They were then met by Roberto Farinacci, the *ras* of Cremona,[17] who together with local and Magneti Marelli Dopolavoro leaders 'took part in paying homage to the Glorious Dead of the Fascist Revolution'.[18]

From 1935 onwards the Magneti Dopolavoro was involved in the mobilization of public support for the war effort. The collection of precious metals to help finance the war was essentially an attempt to make everyone feel that they were personally actively contributing to the military campaign. *Sprazzi e Bagliori* reported on the ceremony, where 1,025 kilos of gold and 15,810 kilos of silver were collected, with its usual sentimental flourish (and possibly a certain disregard for truth):

> Simply to mention the spontaneous eagerness to give of many of the workers and the fact that many deeply regretted that they could only give so little, would still not be a full enough description of what happened . . . Many brought gold and silver articles which had sentimental value and were worth far more than they cost in money to the donors. Yet no one seemed to regret their sacrifice and on every

[16] *SB* 5 (1935), 49.

[17] Farinacci was one of the most powerful local Fascist leaders (*ras*) who emerged in the early days of Fascism's violent rise to power.

[18] *SB* 2 (1937), 50–1.

face could be seen the joy of giving, because everyone is fully aware of the justness of Italy's cause in this war.[19]

The female members of the work-force were given a special chance to do their bit in

the moving and meaningful ceremony of the Widows and Mothers of the War Dead, in which, inspired by the example of the Queen of Italy, a large number of our Dopolavoro members participated and 270 wedding rings were offered. The Party hierarchs kindly allowed us to hold the event on FIMM premises. In a moving ceremony attended by all the Dopolavoro members, the donors received steel rings, 'symbols of affection and faith', from the graceful hands of Donna Paola Quintavalle in exchange for their wedding rings.[20]

Apart from this public sacrifice of their treasured possessions the Dopolavoro members could contribute to the war effort in the sending of parcels to soldiers on the front and their families left behind in Italy. During the Second World War this activity became more organized and the 'Dopolavoro of the Armed Forces' was set up to send relief parcels to the front, arrange special parties for the staff of hospital trains passing through Sesto, and take food parcels to wounded soldiers in hospital.[21] A German course was organized and from time to time patriotic lectures were given and war documentaries shown.

The African war effort seems to have provoked a certain amount of enthusiasm from at least a section of the membership. On one occasion an excess of enthusiasm led to a carnival float made by the Magneti Dopolavoro being banned as inflammatory by the local police. The float was filled with people dressed as donkeys. As the *Sprazzi e Bagliori* commentator wrote, 'Beautifully aptly the carnival float was entitled "the school for donkeys" and featured the fifty nations of the Geneva League sitting on benches in a school where victorious Italy played the teacher's role.'[22] These activities implied a much more active level of political participation than accepting the limited token propaganda included in the various outings, and this was the only period in which the Magneti Dopolavoro (along with the OND in general) really tried to act as an agent of political mass

[19] *SB* 1 (1936), 37. [20] Ibid. [21] *SB* 5 (1940), 40.

[22] 'Relazione sulle attività del Dopolavoro Aziendale Magneti Marelli nell'anno 1937—XV', *SB* 1 (1938), 55.

mobilization, although the extent of its effectiveness is difficult to evaluate.

On the whole, however, the political content of the Dopolavoro activities was much less blatant and probably could be ignored, to a certain extent, by the majority of the participants, who were motivated principally by the desire to have a good time. Sometimes the political aims were, in any case, confusing and possibly the organizers themselves may have been unclear about the exact message they were supposed to be spreading to the masses. This confusion is epitomized by the annual 'Grape Fair'. It was a policy of the OND to revive local 'traditional' fairs (or invent them if necessary) as part of the regime's ruralization policy, and also (as a result of a rather convoluted logic) to promote a national spirit even though the celebrations were very much local. This was sharply in contradiction to the 'technocratic' ideas of the management and indeed of the early years of the OND. It has frequently been noted that Fascist policy was a mass of contradictions, and this is clearly reflected in the absurdity of an annual fair in celebration of the grape harvest with traditional games such as *tamburello* and *piantachiodi* and dancing in rustic costumes all in an effort to nostalgically recapture the flavour of bygone times, organized by Michele Soliveri, dedicated Fascist, but also works manager in one of the most technologically and managerially advanced firms in the whole of inter-war Italy! Perhaps, therefore, the ideological content of these events should not be taken too seriously. Events like the 'Grape Fair' were mostly an excuse for a colourful party, a lively moment in what was a very drab existence for many, and its popularity was undeniable. In 1938 it was claimed that 3,000 persons attended.[23]

Apart from its involvement in the wartime mobilization of the civilian population the Magneti Dopolavoro was not overtly a highly politicized organization. Even its limited political role was at times confused and contradictory, since the needs of the firm and the needs of the regime did not always coincide. Magneti in fact largely solved this dilemma by leaving only the relatively unimportant recreational facilities in the sphere of the Dopolavoro. Thus conflict was minimized. This relative depolitic-

[23] *SB* 6 (1938), 48.

ization reflected the situation in the OND as a whole and this, as Victoria De Grazia has quite rightly argued, was a major factor in its success in recruiting members among the working class.[24] Certainly, going on a day trip with a group of fellow workers can hardly have seemed a great gesture in support of Fascism, even if it was necessary to put up with a bit of Fascist ceremony and speechmaking during the visit. As one ex-worker argued when asked why people went to the Dopolavoro:

> They were people who had never seen anything. They wanted to see something new. And when there was that chance there with the firm putting some money towards it (it really was cheap to go), then everyone, whether they were Fascists or not . . . Of course there was always some of their propaganda involved, there was always someone in uniform who was in charge, who organized the thing. That's true! But there were apolitical people who went too, or they might even have been political but not with the same ideas as them (Ercole)

FROM TENNIS TO 'TAMBURELLO': ACTIVITIES OF THE DOPOLAVORO

The expeditions and outings were by far the most popular activities. The annual trip to Venice, which included a little propaganda but was mostly just sightseeing and relaxation, attracted over a thousand members in 1935 and 1937. Numerous other outings usually on Sundays were made to nearby towns, monuments, factories, or exhibitions. In 1934, for example, 1,520 people went to Genoa to visit the ship 'Transatlantico Rex', and in April 1936 there was a three-day excursion to Umbria, Tuscany, and Romagna. In 1938 there was even a week's cruise from Genoa to Naples during the annual week of holiday.

Sporting activities were emphasized from the 1920s. There were gymnastic sessions for girls and a Sports Group. Skiing expeditions were being organized by 1927 and football by 1930. Benni himself defended the importance of sports for workers against the criticisms of the Fascist unions, bitter at their exclusion from the OND.[25] By 1934 there was more choice of

[24] De Grazia, *Culture of Consent*, 45. [25] Ibid. 105.

sports, including football, canoeing, *tamburello*, motor-cycling, bowls, running, tennis, skiing, and cycling.[26] At other times there was also basketball, gymnastics, billiards, and swimming. Apart from the sports field inaugurated in 1934, a volleyball court and a tennis court were added in 1938 and the canoeing group had its own facilities in Milan from the mid-1930s. According to the factory newspaper, the level of active participation was reasonably high. In 1936, for example, there were 109 members in the rowing section, 203 cyclists, 25 motorcyclists, and 128 bowls players. A fair number of workers did use these facilities, but these figures pale into insignificance beside the fact that in the same year the Excursions Section had 3,749 members![27]

The sports groups were in theory open to all and the financial resources of the members were not supposed to be an obstacle to their participation in the sport of their choice. This was because the Dopolavoro was supposed to be promoting the spirit of class co-operation, one of its main attractions for the management. In late 1928, for example, the factory newspaper wrote gushingly that

> On numerous occasions we have had the pleasure of describing the increasing numbers of trips where workers, clerks, and managers join together to enjoy healthy tourist trips, building bonds of brotherhood in the restful beauty of nature.[28]

In theory this was true of all the sports activities, and *Sprazzi e Bagliori* stressed that

> those of our employees who love motor-bike racing can find in the Dopolavoro not only the inspiration to practise this sport but also *material help* which renders it possible for every type of employee to have a chance to practise a sport which truly cultivates strength and healthy competitiveness.[29] (my italics)

But, in spite of this claim, class barriers remained. Skiing and mountaineering, in particular, were practised almost exclusively by directors, engineers, and a few clerks. The majority of

[26] 'L'attività del Dopolavoro nell'anno XII', *SB* 1 (1935), 99.
[27] 'Relazione sulle attività del Dopolavoro Aziendale Magneti Marelli nell'anno 1936—XIV', *SB* 2 (1937), 40–2.
[28] 'Turismo della FIMM', *SB* 10 (1928), 29.
[29] *SB* 2 (1938), 52.

workers, on the other hand, were left with a choice of more humble sports such as running, football, or bowls. The 'material help' offered by the firm was clearly not sufficient to make expensive sports accessible to the ordinary worker, since even a subsidized motor-cycle or pair of skis cost far too much on a tight budget.

The OND had very specific ideas on which type of sports were useful to the masses. Competitions were to be reserved for an élite of potential champions, whilst the members of the Dopolavoro should concentrate on participating in a friendly way, improving their health, and enjoying themselves. The revival of traditional sports such as bowls and *tamburello* was intended as a contribution to the 'nationalization of the masses' by favouring truly Italian activities rather than foreign imports such as football. The only sign that the Magneti section paid any attention at all to these directives was its inclusion of traditional games and curbing of the activities of the football team in the late 1930s, 'following the orders of the OND which has decided that football is an unsuitable activity for Dopolavoro members'.[30] But these were small concessions, since Magneti, like Fiat, seems to have totally ignored the OND concept of suitable sports and concentrated instead on building up crack teams capable of winning as many trophies as possible, dressed up in the firm's colours. Even potentially non-competitive sports such as gymnastics were seen in this way:

> The young athletes, *chosen* amongst the most suitable women workers, under the expert tuition of Gino Nozzo, have followed a complex training programme and by now have learned to do the most complex group gymnastics. A programme of track running, high jumping, long jumping, and javelin and discus throwing have made it possible *to pick out the most promising women who will now be given special training*.[31] (my italics)

A few of these sporting events provided the occasion for political propaganda, but there was much more stress placed on the wearing of team colours, the trophies, and the training, which were all meant to encourage the workers to identify with the firm. Competitors were made to feel that they represented

[30] 'Relazione sulle attività . . . nell'anno 1937', 55.
[31] *SB* 1 (1939), 40.

their firm and as such were responsible for its reputation. Like Fiat, Magneti seems to have been more interested in co-opting its own work-force than in 'nationalizing the masses'. To a certain extent these aims coincided, but where they did not, the loose structure of the OND made it easy for an individual firm to shape the Dopolavoro to its needs.

There was also a limited programme of 'welfare' initiatives consisting in a small number of loans to workers in distress, subsidies to enable workers' children to go on seaside holidays, and the sale of goods at reduced prices which could be paid for in instalments. The members were apparently enthusiastic about these initiatives, since both the 1936 and 1937 reports note that they had asked for an expansion of the welfare programme, but with the financial crisis of 1937 this was the first thing to be cut. In the words of the annual report:

> We decided it was necessary to recommend the new Dopolavoro Council to limit the funds spent on Welfare Activities, not because we do not understand the noble and humanitarian aims of this work, but because we feel that this should really be done by other organizations which are better equipped for it than the Dopolavoro.[32]

Charitable generosity was only possible while the membership subscriptions were very high. Once these were cut even this limited welfare programme was immediately abandoned, although the instalment buying scheme continued unabated since the cost to the organizers was minimal.

CULTURAL ACTIVITIES

The Magneti workers could also take part in a number of cultural activities. The OND devoted less energy to elevating workers' culture than its counterparts in Nazi Germany, and Tannenbaum has rightly characterized the spirit of the Dopolavoro as 'frankly hedonistic and plebeian'. He considers that the 'mediocre quality of Dopolavoro activities and the lack of opportunities for self-improvement were due as much to the organization's leadership as to the wishes and tastes of its

[32] 'Relazione sulle attività . . . nell'anno 1937', 55.

members'.[33] At Magneti there was a library, but most of the cultural activities of the Dopolavoro were closer to Tannenbaum's description.

The Magneti Dopolavoro had its own band from the early 1930s and a theatre and orchestra from 1938. The theatre was not used for professional companies but for plays acted, directed, and sometimes even written by members of the work-force. According to *Sprazzi e Bagliori*, this was because, although professional actors might produce better plays, 'the Dopolavoro . . . should not be simply transformed into a theatrical business since it was created to educate and amuse the work-force. This kind of show is really what the Dopolavoro should be all about.'[34] The same principle was applied to the concerts held in the theatre, and the first

> grand workers' concert at Magneti Marelli . . . was quite different from those held in other firms because it was not given by a professional group of musicians. Here, by contrast, all the players and singers were authentic workers or clerks, employees of the firm, who had learned voluntarily in the firm's Dopolavoro band or choir.[35]

The plays and concerts produced do not seem to have included any serious music or drama. The first theatre production was a series of musical sketches written by one of the firm's clerks. The music was described as having 'simple tunes'.[36] A few months later the drama group produced *La fidanzata di Cesare* by Silvio Zamboldi, a 'sweet, old-fashioned, sentimental, and cheeky comedy, which brings back our fondest memories of our schooldays'.[37] When concerts were organized the music was 'very Italian and truly popular', mostly songs from light opera with a sprinkling of Fascist tunes like 'Giovinezza', and usually including the specially composed 'Magneti Marelli Dopolavoro Anthem'.[38]

These type of activities seem to have reflected the tastes of the members—over 5,000 people attended the band's first concert. This was not high culture but it was popular and it gave the members a rare chance for some creative activity. In fact the

[33] E. Tannenbaum, *The Fascist Experience: Italian Society and Culture 1922–1945* (New York, 1972), 138. [34] *SB* 3 (1938), 58. [35] Ibid. 55.
[36] Ibid. [37] *SB* 4 (1938), 48. [38] *SB* 3 (1938), 57.

Dopolavoro activities educated the workers and broadened their cultural horizons only to the extent that they gave them a chance to visit new places, practise sports, learn how to act, or play in a band.

The vast majority of Dopolavoro activities were basically hedonistic, giving workers and clerks a pleasurable way to spend a Sunday afternoon or an evening after work. This was largely due to the management's decision to keep the more 'serious' welfare initiatives separate, but it also reflected the aims of the OND in the 1930s: to create a mass 'consensus' for the regime even at the cost of diluting the political content.

WOMEN AND LEISURE

The question of women's participation in the Dopolavoro is of special interest in view of the arguments of certain historians that factory OND sections were the means of mass socialization into the nation of the working class. According to Tannenbaum, for example, it was the OND, rather than the regime itself, which gave workers 'the beginnings of a feeling of identification with a modern institution outside their family or town'.[39] This may be true for men, but it is far less accurate in describing the impact of the Dopolavoro on female workers. A study of their lives outside work hours suggests the irrelevance of this institution to this half of the working class and lends weight to the argument that the use of the concept of 'leisure' is highly problematic in looking at the non-waged work activities of women, since 'women's position in waged and unwaged work in relation to the family means that the existence of leisure as a pure category for women is questionable'.[40]

In contrast to the provisions for the male workers the Magneti Dopolavoro offered a much more limited range of facilities for the female members of the work-force. Women did take part in some of the activities and there were, for example, a few female names in the programmes of the concerts and theatrical productions, but usually they were singers or dancers and only

[39] Tannenbaum, *The Fascist Experience*, 139.
[40] C. Griffen *et al.*, 'Women and Leisure', in J. Hargreaves (ed.), *Sport, Culture and Ideology* (London, 1982), 88.

very rarely did women play musical instruments. Larger numbers of women almost certainly went on the excursions, but unfortunately the figures available for these trips are not differentiated by sex.

Women took part in some sports, but the numbers of men involved was always far greater. Men played football, rowed, cycled, and rode motor-bikes, to name but a few of their sporting activities, whereas the women seem to have been limited to athletics, gymnastics, cycling, and volleyball. In fact only certain sports were encouraged for women, those that were seen as suitable because they would not make the women too masculine. This reflected the contradictory attitude the Fascists had towards women's sports. In the 1920s women's sport (which had been very limited in the pre-Fascist period) was encouraged because it was thought to make women healthier childbearers. But even in this early period and increasingly in the 1930s many leading Fascists (like the Vatican) voiced fears that women's sport could lead to immodesty and immorality, partly because of the type of dress involved.[41] The Magneti factory newspaper seems to have shared these anxieties. In 1933 it stressed:

> In our opinion, there is one infallible rule in deciding which games are suitable for women: we would exclude those which require those movements, efforts, expressions, or clothes which offend good taste and morality . . . How painful it is to see feminine faces wasted and almost ruined by exertion, their clothes bathed in sweat, in short a total destruction of all that can be summed up in the single word—femininity.[42]

Concern with the preservation of femininity was not limited to sports activities. The firm's Dopolavoro also organized special all-female outings. Here the emphasis was on respectability,

[41] One good example of this contradictory attitude was when a group of women in Milan (with a male trainer) obtained permission from the Fascist authorities to set up what was probably the first ever women's football club in Italy. Permission was only given on condition that training and matches should take place in private, in other words on enclosed pitches without public spectators. (S. Giuntini, 'La donna e lo sport in Lombardia durante il fascismo', paper given at conference on 'Donna lombarda 1860–1945' (Milan, 1989).)

[42] 'La donna e lo sport', *SB* 2 (1933), 5. It is worth noting, nevertheless, that in practice women did, in fact, occasionally do more strenuous sports, such as discus throwing.

religion, and patriotism. These trips were usually for the women from the Winding Shop, accompanied by their head of shop. The following description of one outing gives a flavour of how the firm liked to view these occasions:

> The outing . . . which was for female employees only, took place on the 8th of December. A large number of women, led by Signora Luigia Pedroni, a member of the Council of our Dopolavoro and the organizer of the event, made a devout pilgrimage to the Sanctuary of the Madonna of Caravaggio. Amusement and enjoyment were put aside for a while, in favour of a more religious atmosphere and they raised their voices in prayers for our beloved distant soldiers and for the victory of Italy in war.[43]

Apart from outings, there was one specific 'leisure' initiative which was aimed exclusively at the female members of the work-force. This was the dressmaking and domestic science school. Founded in 1924, the school is frequently referred to in the factory newspaper as a Dopolavoro initiative, but in fact the courses seem to have been run by the factory social worker, Evelina Garbagnati. The school, one of a number set up in Italian firms in this period,[44] was designed to help reconcile the employers' desire to employ cheap female labour with Fascist ideology stressing that women's place was in the home. Domestic science courses were seen as necessary to ensure that women would not become 'unfeminized' by factory work. It was felt that they needed to be taught how to do housework, a craft that they would otherwise learn 'automatically' if they did not go out to work.[45] For the women themselves this was supposed to give them a chance to contribute to their families; a thrifty housewife could stretch a tight budget by good and economical housekeeping, and by making clothes for herself and her family. On the shoulders of the women, therefore, rested the responsibility for the survival of the household faced

[43] 'Con il Dopolavoro Magneti Marelli', *SB* 1 (1936), 39.

[44] Some of the domestic science schools for female workers in Milan were set up under the auspices of the women's Fascist groups (*fasci femminili*) whose Dopolavoro section was headed by none other than Giuseppina Benni, wife of Antonio Stefano Benni, the Magneti president.

[45] This was explicit in Fascist writings on the subject. See, for example, the preface by Fascist Party Secretary A. Turati to the volume of collected papers from the 4th International Domestic Science Conference, A. Turati, 'Per le scuole di economia domestica nell'industria', ASNI, 5 (1928), 2.

with wage cuts and unemployment in periods of economic crisis.[46] The firm even liked to stress how dressmaking could be a source of income for workers at these times:

All of these women learn in this school much which will be of use to them later in their family lives when they become wives and mothers. Some who are particularly good, furthermore, make use of their skills outside the school by doing a little tailoring or embroidery, at least in those lean periods when they are temporarily laid off.[47]

Unlike male leisure pursuits which were more pleasure-orientated, women's 'leisure' as provided by the Dopolavoro was seen as intrinsically bound up with duty to home and family and defined by the prevailing norms of domesticity and femininity. The emphasis was on duty and sacrifice. The domestic science school at Magneti reflected all these concerns, as is evident in the following extract from an article written by Garbagnati addressed to the women and girls registering for the course in 1928:

The kind of teaching you will receive aims above all *to put you women in your proper place*. The necessities of life today with the increased industrial expansion of the last few years have meant that women too have been called to work in factories, tearing them away from *their true mission: the care of the home, the protection of the domestic realm*.

The decision to set up this school was taken by the Magneti Marelli management because, as you can see, my dear friends, they had realized how much your knowledge of domestic matters was lacking and feared that young girls might forget their real mission in life. They felt that in the education of every good woman worker there should also be included some training in how to be a good housewife . . . A wise and clever woman needs to make use of all her knowledge. It will

[46] e.g. an article which appeared in ASNI in 1932, describing the dressmaking school in the Ercole Marelli factory, stressed that these initiatives were important and that 'more women should attend the school . . . it would be in their own interests. They can learn everything there, from cutting out garments to making them up. Mothers and daughters should reflect carefully on the amount of money their households could save if they all attended these wonderful schools. The entire expenditure in the family budget for women's clothing could be saved. Mothers should really work hard to convince their daughters to attend the school. Let us not forget that it is these type of expenses, more than food, which eat away at hard-earned pay-packets.' (P. Preda, 'Una visita alle opere assistenziali della società Ercole Marelli', ASNI, 5 (1932), 84.)

[47] 'Una ceremonia gentile alla Magneti Marelli', *SB* 4 (1938), 46.

represent an indestructible heritage to her family which will profit from her knowledge of economical budgeting.

And since a life of comfort is based on the twin benefits of work and thrift, the more we can save money, the more likely it is that in our homes there will be prosperity and well-being.[48] (my italics)

Apart from these special courses, however, the evidence provided by my interviews with female ex-Magneti workers suggests that women's involvement in the Magneti Dopolavoro was minimal. Since membership was effectively compulsory it seems safe to assume that most joined. Few, however, actually seem to have taken part. Only a couple of the women I interviewed remembered participating in any of the Dopolavoro activities, and then only in a sporadic way. Most of them seemed barely aware of its existence and had only the vaguest memories of its activities. A few had attended the dressmaking classes but others had never even heard of such a thing and, in fact, some of the interviewees declared that they did not think that there actually was a 'real' Dopolavoro at all before the war. One of them, for example, in many ways more aware of the factory as a whole than many of the other interviewees due to her position as an overlooker, said that the Dopolavoro was simply 'a place for recreation. They used to go there to play cards, the girls maybe . . . they went there out of friendship' (Orsola).

This contrasted with the men, most of whom were able to give a good account of what the Dopolavoro provided. For example, Ercole (who himself went to the Dopolavoro canoe club) said 'at the Dopolavoro you could play bowls or ball games. There was basketball. There was there, in front, a lovely athletics field . . . and so on.' Another male ex-worker, Giuseppe R., also remembered the Dopolavoro well:

There was a pretty extensive, pretty sociable free time organization, organized by the firm. In the first place it was there because the firm was doing well, they were making a lot of profit, eh? And they had set up . . . football teams, basketball teams . . . We had football and other things. I loved football.

The majority of the female interviewees were totally unable to recollect details like this. It is also revealing to note exactly

[48] E. Garbagnati, 'Corso di economia domestica 1928–29', *SB* 11 (1928), 41.

which aspects of factory life were forgotten by the women and which were remembered. A number of women could remember little or nothing about the Dopolavoro, whereas every single woman I interviewed, even the childless, clearly recalled the nursery. This suggests that few women actually got involved in the Dopolavoro in any way. Some of the women, furthermore, commented that the Dopolavoro was mainly for men, and one thought that the firm provided nothing for women at all except the choir.

The limited participation of women may be explained in part at least by the fact that a good proportion of them were employed in the strictly disciplined Winding Shop, where the timetable made it hard to get involved in after-work activities. One of the ex-winders explained that she went to the Dopolavoro for the first time in 1943 when they began working a more normal timetable:

> I went there many times in the '43 to '45 period. Yes, yes, I used to go there once we had started to work normal shifts. But before then I couldn't go. Our timetable was no use for going to the Dopolavoro. (Maria V.)

The explanation for the low level of participation in the Dopolavoro is, however, more complex than the particular rules of one shop, since it was true of all the Magneti women. It also reflected the very low rate of participation of women in the OND as a whole. In 1930, the last date for which separate figures for men and women were kept, a mere 10 per cent of the national membership was female.[49] To understand why participation was so low, at least in the case of the Magneti women, it is necessary to consider their home lives, which gave them little freedom.

There were two main constraints on women's freedom in their home lives. One was the fact that after work they usually had to do housework, thereby severely limiting their free time after work. The other was the sexual double standard, which meant that girls and women were strictly watched by their parents and husbands for behaviour which might endanger their reputations.

[49] De Grazia, *Culture of Consent*, 43 n.

HOUSEWORK

Housework was heavy in this period, and particularly so in the cramped and overcrowded housing of the urban working class.[50] Frequently whole families crowded together into only one or two small rooms. In some cases water had to be fetched from outside, and domestic appliances, which were beginning to be available to the working classes in certain other countries such as Britain,[51] were only for the wealthy in Italy.[52] Some housing had gas lighting by this time, but heating was almost invariably by wood- or coal-burning stoves, which required a good deal of time-consuming care.

Most girls were expected to do a fair amount of housework. Once they had worked a full day in the factory and then helped clean the house there was little time left for joining the factory sports clubs. As girls in the parental home, the contribution that they were expected to make varied considerably, depending on factors such as the size of the family, the length of their working day, and whether or not their mothers were in full-time employment. If their mothers were at home some did little: 'And what could I have done? I used to help a little on Saturdays and on Sundays when I was at home. After all my mother was at home, she used to cook, she did everything, what could I have done?' (Gina). Delfina was expected to do nothing at home, and commented 'I learned once I got married'. Others said they had no time to help at home 'because I was working and did ten hours a day' (Germana). The length of the working day did, indeed, mean that many of these women had practically no free time at all. For Maria V., who came from a rural village, the day's timetable was particularly gruelling: 'I used to go out of the house at half past three (in the morning) and I didn't get home until quarter to five or ten to five (in the evening).' Yet

[50] On the development of housing in Milan see G. Consonni and G. Tonon, 'Aspetti della questione urbana a Milano dal fascismo alla ricostruzione', in *Classe*, 12 (1976), 43–100; Consonni and Tonon, 'Milano, classe e metropoli tra due economie di guerra', *Annali Feltrinelli* (1979–80).

[51] See J. Lewis, *Women in England, 1870–1950* (Brighton, 1984), 151–2.

[52] C. Saraceno, 'La famiglia operaia sotto il fascismo', *Annali Feltrinelli* (1979–80).

even some girls who worked very long hours had to help with the housework, particularly if their mothers worked:

[Q. What kinds of housework did you have to do?] Washing, ironing, etcetera. I had two brothers at home until the one who's dead now got married. My father worked and my mother had to help him too so . . . (Giuseppina)

A few boys did do some housework. Giglio, for example, when 12 years old had to look after his baby sister while his parents were out working in the fields. Ercole said he had to help too:

I did housework on Saturdays. We used to wax the floor . . . We polished, we cleaned . . . the furniture they used to have in those days was all inlaid, it was hard work . . . dusty. I always had to help too.

But even here the slightly older sister seems to have done much more, and it was she who took over the cooking and most of the housework at about the age of 10 as the grandmother who up until then had done most of the work was getting old.

Boys like Ercole and Giglio were somewhat exceptional in that they had to help at all in the house. More typical was Luigi B., who, when asked if he did housework as a boy, replied in a shocked voice, 'No! The household tasks were done by the women.' Pietro gave the same view when he said that he had only to fetch water from the spring as a boy in a rural area and 'if it was to do with the house my sisters had to do it'. Giuseppe R. did no housework when he lived with his family, saying that housework was for girls:

I wasn't good at it . . . But seeing as I went to work, I was working and I was studying too. There was no chance I'd have enough time to do anything else. I didn't even have any free time . . . My sister did as much as my mother . . . Of course . . . Because with four male children and only her . . . eh?

The situation in Giuseppe R.'s family seems to have been typical. Boys generally only did household chores in exceptional circumstances where no girl or woman was available to do the work. Where children did have to help it was seen as primarily work for daughters and not sons. Pierina recalled:

we were four sisters and we used to split up the work between us: one did the cleaning, another did the ironing, another the washing . . . we divided everything up . . . the housework was done by us women rather by them (my brothers), because they used to work in the factory, then they had to help my father, then they were free. We girls on the other hand had to go to work in the factory, then we had to help our father on the land, and then after all that we still had to wash, iron, do the housework, and generally help our mother.

It was with marriage, however, that the real burden began. Some husbands did nothing at all in the house. One man I interviewed boasted that all his life he had never had to do any housework at all: 'Me? I don't know how to do anything . . . not even make coffee . . . even nowadays' (Giuseppe). This man's wife worked at home as a dressmaker, but the women who worked full-time at Magneti often had just as little help. Luciana said 'No, no, I did everything, men aren't used to helping.' Other women made similar remarks:

He did absolutely nothing at all, not even move his own chair backwards and forwards, I did absolutely everything. (Germana)

What do you expect? A man's a man for all that. [Q. And didn't you ever complain about this state of affairs?] Sometimes yes, I used to say—you could at least tidy up! He did learn a bit, he started putting his own shoes and slippers away. (Orsola)

Not all men were like this, however. Some couples seem to have enjoyed a more equal and sharing relationship:

He helped me a lot: he used to do the shopping, and that means a lot and sometimes I came home and the food was ready. If he was on the first shift he used to get home before me, if he was on the late shift I was first . . . he didn't complain, I really mean it. (Carla)

If I got home late he used to prepare the food and if I got home first then I used to do it . . . He also used to get the child up in the morning and dress her and take her to my mother's, because I had to go out first. He always made the bed . . . he kept the house clean and tidy. I used to start in the evening and he finished off in the morning . . . We both had to go to work so there was no distinction between husband and wife, there was only the difference between who had the time and who didn't. (Maria G.)

He used to help me. I used to get in late and he would have already started preparing. Yes, as far as that goes, he used to help me. (Olga)

Pietro, who as a boy had done no housework, began to help

when he was married in the city: 'When we were there I did [housework] too . . . Yes more than once, cleaned the floor, cleaned the windows, cooked . . . Whoever got in first got the stove going.' In all of these cases, however, this was not a real equal sharing of the housework, since none of the husbands did the washing of clothes, easily the biggest and most time-consuming task, and the women almost always had the responsibility of the family budgeting: 'No, he only used to help me with the cooking, lighting the stove, and other little things like that . . . When I got home I still had all the washing to do' (Olga). Thus even a husband who did some housework was no more than a helper, secondary to the wife who did the bulk of the chores after her day in the factory. Nevertheless, this evidence that some men did do a certain amount of housework does suggest that at least to a limited extent industrialization and factory work could lead to a lessening of rigid gender roles in the home.

RESPECTABILITY

Another factor which severely constrained women's leisure activities was the rigid sexual double standard which meant that 'respectable' girls and women rarely went out. As young girls the women that I interviewed had very little freedom to go out either alone or in the company of friends. As Orsola commented, 'there was a lot of difference between women and men in those days because we weren't allowed to go out'. Men were able to reminisce about their youthful freedom. Giglio, for example, remembered: 'Yes. I used to go out with a friend. We used to go to Milan, go to something like the cinema or to see a show.' Giuseppe R.'s comments were particularly revealing: 'Milan was one of the cities in which there was a lot of freedom. It wasn't dull like in the countryside where everyone stays at home.' But Giuseppe R. was forced to admit that this was only for men. When questioned if his sister enjoyed this freedom he admitted: 'No! My sister never went out . . . It was our parents who wouldn't let her go out. At the very most she was allowed to stay out until nine o'clock.'

Most women, like Giuseppe R.'s sister, did not share in the 'freedom' afforded by Milan. They, in contrast to the men, were usually allowed to go out only on Sunday afternoons or very rarely in the evenings, and generally only if accompanied by a relative and to certain places: 'I wasn't allowed out, I only started going out once I'd got to know my husband, and even then they sometimes used to tell me off: I was 23 years old and they still told me off!' (Pierina). Some women enjoyed a certain amount of freedom. Germana, for example, was allowed to go dancing with her female cousin in the afternoons:

> I lived there, and so I was friendly with lots of girls around there, you know what it's like, when you're young you make friends easily. And we used to go out dancing together in a gang and in the evening we came back home together. That's how I got to know my husband.

But this was fairly unusual, and most others were more strictly watched by their parents: 'oh my goodness, my father was so jealous of me, I was his first daughter. He'd had three sons before me and when I was born . . . he'd wanted a girl so much . . . it was as if the sun had come out!' (Orsola).

In retrospect most of the women thought the limitations on their freedom had been a bit too strict:

> In those days it wasn't like now . . . If I'm with some friends, we just go to a bar and drink a cup of coffee. Nowadays a woman . . . for example if I'm in town and I haven't had a coffee yet, I just go into a bar and have one. In those days you couldn't do anything like that. It was partly the way our parents had brought us up, you know what I mean? We were pretty backward in those days. (Giuseppina)

This comment, however, clearly reflects the intervening years during which Italian women have enjoyed far more freedom. So it is difficult to tell what these women felt at the time, whether they accepted or questioned these restraints. One woman recounted how she had had to pretend to be going out with her brother in order to see her boyfriend, but on the whole women were probably unwilling to defy their parents and risk getting a bad reputation, thereby endangering their chances of getting married.

WOMEN'S LEISURE ACTIVITIES

What then did these women actually do in the limited free time available to them after they had finished factory work and housework? For many women, especially if they lived in rural areas, the church was still more important than the Fascist Dopolavoro as a focus for social life. A number of women did in fact say that they went to their local church's youth club (*oratorio*), this being seen as one of the few respectable ways for a girl to go out of the house. One woman, for example, said of Magneti's facilities that 'there wasn't a real Dopolavoro like the sort of thing the church organized' (Orsola). In rural areas particularly the Fascist Dopolavoro had penetrated little[53] and the church had maintained its pre-eminence as a provider of free-time activities.[54] For men there was always the local bar, but this was not respectable for women. Where no leisure facilities existed at all, simply going to mass became an important social activity. Even for some women in more urbanized areas the church often still remained an important centre for socializing. As Pierina remembered: 'But in the evenings we didn't go out, if we went out at all it would be to a religious procession or to church but never to enjoy ourselves.'

Apart from church-going most of the female interviewees did refer to other leisure pursuits. In general, activities which can be seen as purely for pleasure were rare. One woman describes how she used to go cycling:

> together with friends from work, yes. I especially loved going out for a nice cycle trip on a Sunday. Three, four, or five of us used to go, all women, or there might be a man with us, someone's boyfriend, and off we went . . . In the cherry season we used to go out to Brianza on bikes, we used to go off in the morning and come home in the evening. Sometimes I could leave my children with my mother . . . Even after I got married, even when I was with my second husband, every now and then we used to go out on a little trip together. (Olga)

[53] In the agricultural areas of the province of Milan the percentage of the population who belonged to the OND was much smaller than in industrialized zones. In 1938, for example, 26% of the population of Monza (an industrial town) were in the OND. In some agricultural rural areas the percentage was as low as 14% (Lodi) or even 6% (Casalpusterlengo) (F. Tintorri, 'L'OND a Milano (1923–1939)', paper given at conference on 'Industria, fabbrica, territorio nella Lombardia degli anni '30' (Milan, 1984)).

[54] The Fascist state allowed the Catholic Church to keep its mass organization *Azione Cattolica* (Catholic Action) and its women's circles. It had to agree, however, to the dissolution of the Catholic Sports Federation.

But Olga was unusual. Most of the other women had no memories of such carefree moments, although some of them remembered going dancing: 'To tell you the truth, I just loved dancing, because we didn't have any entertainment, nothing but work, work, work' (Germana). Yet even this was not purely for pleasure, as girls saw the dance hall as a good place to meet a potential husband and secure their future. A number of the women that I interviewed met their husbands at afternoon dances or at the cinema.

Many women also sewed in their free time, and a number of interviewees described this enthusiastically as their most important recreational activity.[55] Delfina said 'I always loved sewing.' Orsola remembered, 'I was crazy about sewing. Do you realize that I've always made all my own clothes, even now.' These women's love of needlework was obviously important to them. Sewing was not, however, simply a hedonistic activity, the female equivalent of the men's night out in a bar, since it was also useful and productive and as such the epitome of female 'leisure' activities:

> I made my entire trousseau for when I got married, and I loved sewing and so I used to go home and start embroidering . . . So I used to fill up my spare time that way. Because in those days we used to make our own trousseau. We didn't have anything made. Everything was made by hand, the pillowslips . . . we made everything we needed. (Delfina)

> I had that teacher of mine there. So when I had a couple of free hours of an evening (I always had so much work to do) she used to say to me 'Nina, if you've got time this evening . . .' I used to go and help her. Then if I had a bit of money I used to buy a bit of cloth and she used to make me my dresses. That way I always had plenty of different things to wear. (Delfina)

Sometimes this could even become a source of extra income. Germana, for example, recalled that, when she was not too tired after work,

> I used to stay up if I had some knitting to do. I made all of my own things myself and I used to work for a woman as well. For the girls at Magneti too, because the women there used to say to me 'Can you

[55] On needlework as a female 'leisure' activity, see R. Parker, *The Subversive Stitch: Embroidery and the Making of the Feminine* (London, 1984).

make me a nice waistcoat?' You know how it is. And I used to make baby things too, clothes for little girls . . .

In my house there's always been enough to eat, partly because I used to make dresses, because I'd been to sewing school when I'd finished in the factory, in the evening. (Maria G.)

Was sewing, therefore, a 'leisure' activity? It could be an essential part of the household budget, enable women to dress far more elegantly than their wages would otherwise allow, prepare them for marriage, and even become a future career if they eventually had to leave waged employment outside the home because of child care. None the less, as Michelle Abendstern has argued, women's leisure activities should not only be looked at in terms of social control, for they can also be a means of self-expression:

Working class women experience leisure within a dominant culture which is hostile to them both as women and as members of the working class. Their cultural practices reflect this, incorporating, often at the same time and in the same activity, expressions of both acquiescence and resistance, compensation and preference, creativity and repression. It is, and was, a culture of contradictions.[56]

Needlework is a good example of this. Although it embodied all the cultural norms of suitable leisure pursuits for women, it was also a real female craft and a source of pride and interest in a world that gave very little scope for female creativity. It required both ability and imagination and these women saw it as a pleasurable activity rather than simply duty like housework. During her week of annual holiday, for example, Pierina did not go away but instead 'we used to all meet, me and my female friends, and we spent the whole day embroidering'.

Leisure, in the sense of pure recreation distinct from work, barely existed for these women. Their choices were severely limited by parental discipline, long hours of work, housework, and the sexual double standard. Apart from perhaps the odd outing with the firm's Dopolavoro and the occasional visit to the cinema, dance hall, or possibly even an afternoon's cycling, leisure was virtually absent from their lives. In fact it was

[56] M. Abendstern, 'Expression and Control: A Study of Working Class Leisure and Gender, 1918–1939. A Case Study of Rochdale Using Oral History Methods', Ph.D. thesis (Essex, 1986), 49.

unusual enough for Maria V. to remember quite clearly a stolen day, the only day off she ever really had in her work career, when, defying prejudice and parental authority, she and a number of friends escaped for a few brief heady hours of freedom:

> One day we were temporarily laid off but we pretended to go to work just the same and we went to the airfield . . . I said 'Let's go anyway, let's pretend to go to work' and we were out enjoying ourselves all day! We even went to the cinema to see 'Giulia'. We did all sorts of things that day. We almost wet ourselves we laughed so much! What a day . . . Just a gang of girls. (Maria V.)

Incidents like this were, however, rare. Most social life was far more mundane. In spite of the Fascist campaign to 'nationalize the masses', women's focus for social activity remained the home, family, church, and near neighbourhood.

WOMEN AND THE 'NATIONALIZATION OF THE MASSES'

Magneti Marelli offered its workers a wide range of leisure activities and facilities by the 1930s. This was partly a result of the management's desire to emphasize their support for Fascism, and also due to their policy of trying to create a docile and co-operative work-force. The expansion of the Dopolavoro may also have been influenced by the desire of the Fascist unions to create a sphere of influence for themselves. Fascist propaganda was a feature of the Dopolavoro-organized events, but they were probably not seen as particularly politicized by the participants, which accounts—at least in part—for their undoubted popularity.

The testimonies of female ex-employees, however, suggest that they remained largely excluded from this shop-floor culture, and the Fascists' campaign to 'nationalize the masses' through the OND had little impact on this half of the population. Their testimonies, furthermore, suggest that the concept of leisure may be to some extent irrelevant to understanding the lives of working-class women in this period. When waged work in the factory was finished women did not generally rest and relax or study, as some but not all of their brothers and husbands could. Many women's 'non-wage-earning time' was absorbed by duties such as housework. Others were luckier and had a little free time, but even for them the content of their 'leisure' activities was strictly defined by cultural norms which emphasized femininity, domesticity, respectability, and religion. The recreational facilities provided by the firm were largely unused by women, and where the firm did provide specific initiatives for female employees they were designed only to reinforce further the domestically orientated view of women's role.

In short, despite some limited evidence of the emancipatory role of factory work for women in terms of the lessening of rigidly defined gender roles, their lives in this period remained dominated by the concerns and needs of home and family. These limitations on women's ability to participate in the public sphere inevitably had an impact on the extent to which they took part in political activities, whether Fascist or antifascist.

10

'I Was No Hero': Political Attitudes and the Resistance

IN 1943 the Allied forces invaded Sicily and Mussolini fell, deposed in an internal coup by fellow top Fascists who believed that they could save their own skins and their position by ridding themselves of their leader. But the new leadership proved inept and were unable to prevent Mussolini escaping to the north to set up a new Fascist state, little more than a puppet of Nazi Germany, the *Repubblica di Salò*. They also failed to prevent the German army invading much of the Italian peninsula, thereby considerably prolonging the war for the Italian people. Even before the fall of Mussolini the antifascist resistance movement had begun to organize, and for a brief period ('the forty-five days') they emerged from hiding in the summer of 1943. This interlude of comparative freedom ended with the arrival of the Nazi troops in September, but the struggle went on and grew greatly, now directed at the invaders and their allies the *repubblichini*.[1] It continued throughout the eighteen months that were to elapse before the Allied troops finally arrived in northern cities like Milan.[2]

The contribution of the people of Sesto San Giovanni to the resistance movement has become almost legendary.[3] In this period they were, in many respects, the leaders in the Milanese area. As one, often cited, Communist Party report argued in 1944:

[1] The *repubblichini* were the supporters of the Repubblica di Salò. In other words they were the Fascists of the 1943–5 period.

[2] A huge number of books and articles have been published on the history of the Italian Resistance. For a fairly recent overview of the period in English see D. Ellwood, *Italy 1943–45* (Leicester, 1985).

[3] On the Sesto Resistance see *Sesto San Giovanni nella Resistenza* (Sesto San Giovanni, 1973); G. Petrillo and G. Vignati, *Città e fabbrica nella Resistenza, Sesto San Giovanni 1943–1945/documenti* (Sesto San Giovanni, 1975); G. Vignati (ed.), *I ribelli al governo della città: Sesto San Giovanni 1944–1946* (Milan, 1988).

It is undeniable that from every point of view this is our strongest area. This is the zone where our Party is the most influential, our membership is highest and we have carried out the greatest number of actions, we have collected the most money, etc. . . .

This is the key zone to which all the workers of Milan look before entering the struggle.[4]

There are numerous other accounts which portray Sesto in much the same way. To quote one example, the minutes of the first congress of all the Sesto factory CLNs (National Liberation Committees), held in July 1945, noted that:

Sesto is a political and trade union centre which leads Lombardy in terms of activity, demands, and problems of all kinds.

During the months of clandestine activity it was always Sesto that led the way in actions, demonstrations, strikes, etc.[5]

The sheer numbers of people involved in the antifascist struggle give some indication of the magnitude of the Resistance movement in this area. There were over 3,000 Resistance fighters,[6] about a thousand arrests, and hundreds of deportations in the period of Nazi occupation. Resistance activity was strong both in the work-place, particularly the large Breda and Pirelli firms, and also in the town itself. All the different clandestine parties and groupings came to be represented, although the Communists predominated heavily.

The roots of Sesto's left-wing political tradition, however, date back much earlier than this to the turn of the century. Although recruitment to the workers' movement had been initially slow as the engineering sector had tended to be dominated by craft unions with narrow sectorial interests, a broader-based movement began to develop with the foundation of the town-based self-improvement society the *Circolo Avvenire* in 1905 and the foundation of a local Socialist Party branch in 1911. In the local elections of July 1914 with virtual universal

[4] 'G' (Gillo Pontecorvo), 'Rapporto sulla situazione organizzativa della federazione milanese', Milan, 20 Apr. 1944. Original document in APCI. Also reprinted in A. Scalpelli, *Scioperi e guerriglia in Val Padano* (Urbino, 1972).

[5] 'Relazione del primo congresso dei CLN di Sesto San Giovanni', 13 July 1945, ACLNL, b. 4 (Presidenza). This document is partially reproduced in Petrillo and Vignati, *Città e fabbrica nella Resistenza*.

[6] F. Bassoli, 'Alla ricerca di un'identità: per affrontare la sfida del futuro', in Vignati, *I ribelli al governo della città*, 19.

male suffrage the Socialists polled 43.9 per cent of the vote. By 1919 the various workers' circles grouped about 3,000 members and a new Labour Chamber was set up. By the end of the same year Sesto's local administration was in the hands of the Socialists. At the elections of November 1919 they had obtained a massive 66.2 per cent of the vote, and even after the failure of the 'occupation of the factories' they remained strong in the local elections of late 1920. In 1921, following the national split in the Socialist Party, about a hundred, prevalently young, Sesto socialists joined the new Communist Party.[7]

This political tradition of the Sesto workers made them difficult converts for Mussolini's followers, and although the Sesto Fascist group was founded in 1920,[8] it was only in 1925 that they managed to set up a local party branch headquarters.[9] Although tactics of violence were used here as elsewhere (albeit rather later than in some areas), such as the attacks by squads from Milan on the Labour Chamber, the *Circolo Avvenire,* and the Socialist Circle in April and May 1921, this was not enough. As Donald Bell has argued, in Sesto

> the Fascists were confronted by the problem of how to build a consensus and mass base in a mainly working-class environment . . . in the face of a hostile majority, Fascism in Sesto often attempted to disarm and convert its adversaries rather than to defeat them directly. Much time passed before Sesto's Fascists—part of a national movement which focused largely on the violent destruction of opposition—resorted to such violence themselves. This policy, in fact, reflected the special challenge which confronted Fascism in a major industrial location—an environment in which the movement did not find an abundance of the sorts of allies it often had in rural zones and provincial towns.[10]

[7] On Sesto politics in the pre-Fascist period see G. De Vecchi, 'Origini e prima organizzazione del movimento operaio e democratico', in G. Petrillo (ed.), *La città delle fabbriche: Sesto San Giovanni 1880–1945* (Milan, 1981); D. Bell, 'Worker Culture and Worker Politics: The Experience of an Italian Town, 1880–1915', in *Social History*, 3 (1978); Bell, *Sesto San Giovanni: Workers, Culture and Politics in an Italian Town, 1880–1922* (New Brunswick, NJ, 1986), 43–5.

[8] The leader, until his departure for Rome in 1924, was Asvero Gravelli, a worker who was the son of an interventionist socialist. On Gravelli and the difficult rise of Fascism in Sesto see L. Longaretti, 'Il fascio di Sesto San Giovanni: storia delle camicie nere in un centro industriale 1919–1929', tesi di laurea (Milan, 1983–4).

[9] *Sesto San Giovanni nella Resistenza*, 7.

[10] D. Bell, 'Working-Class Culture and Fascism in an Italian Industrial Town, 1918–22', *Social History*, 1 (1984), 12.

Even these flexible tactics met with only limited success, and, although obviously extremely simplistic as a description of the political attitudes of the population of a whole town, the assertion, which appeared in a celebratory publication commemorating the Sesto Resistance, that 'Sesto was always "red", even in the darkest moments of the Fascist regime', is not totally unfounded. Despite Fascist repression, antifascist activity in Sesto, however small at times, persisted throughout the inter-war period.[11] In 1924, at the elections which were won nationally by the Fascists and their allies largely by means of violence, intimidation, and corruption, the Sesto Fascist list obtained far less votes than the parties of the left.[12] Even in 1933, when the Fascist government had already been firmly entrenched in power for a number of years, a Fascist union document complained that

> the Fascist penetration amongst the workers of Sesto seems to be happening only very slowly . . . We have also been informed that in practice our propaganda is sparse and inefficient in the various factories.[13]

But the industrial workers of Sesto did not form the homogeneous mass that these comments suggest. Even in 'Red Sesto' political attitudes and activism varied, for although many of Sesto's workers worked in large firms, their experience depended on many factors; most particularly their gender, level of skill, and influences external to the factory such as the family.

FASCISTS AT MAGNETI

How did Magneti, in many ways so different from other Sesto firms, fit into this picture of antifascist 'Red Sesto'? That the management were closely implicated in and beneficiaries of the

[11] See G. Vignati, 'Prologo: "Il cancro della Lombardia". Antifascisti, ribelli e sabotatori', in Vignati, *I ribelli al governo della città*.

[12] The Fascists won a mere 885 votes compared with 639 for the Unitarian Socialists, 723 for the Maximalist Socialists, 554 for the Communists, and 885 for the Catholic Popular Party.

[13] 'Attività della Unione Sindacati Fascista Industria di Milano fra gli operai di Sesto San Giovanni', Milan, 1933, ASM, Gab. Pref., cart. 1069, f. 'Unione sindacati fascisti dell'industria'.

Fascist regime seems clear. Attempting to assess the political views of their work-force is more complex. The whole issue of whether a 'consensus' for Mussolini's regime ever developed among the working class has proved an extremely thorny issue for historians, and has given rise to some heated controversies.[14] One problem is the question of what exactly should be understood by describing someone as a Fascist. There could, of course, be varying levels of support for Fascism, ranging from participating in a Dopolavoro activity that was largely recreational but included a little propaganda, to being a committed and active party member. In the case of the Magneti workers the evidence on the political attitudes of the work-force is fragmentary and difficult to evaluate, and the only thing that is certain is that a minority of the work-force were political activists, both Fascists and antifascists.

The management frequently attempted to depict the work-force as particularly Fascist. After Mussolini's visit to the firm in 1930 his reception was described as highly enthusiastic:

> There was a continuous commotion while the Duce was visiting the various shops of the factory. It was impossible to hold back the workers who wanted to get close to him, see him, and follow him. When He began to speak from the specially constructed stage, amidst the cries of the crowd there was an almost audible response to the Duce's speech.

[14] The question of whether or not a 'consensus' for the Fascist regime ever emerged among the working class has been the subject of a huge debate among historians. Up until the 1970s historians of the working class in the Fascist period generally avoided this issue, but with the publication of Palmiro Togliatti's *Lezioni sul fascismo* in 1970 some began to question the idea that the working class had been solidly antifascist throughout this period. The debate really took off when historian Renzo de Felice argued that there was a substantial 'consensus' (a term borrowed from functionalist sociology) for the regime in the early 1930s which even extended to the working class. De Felice's work provoked a storm of criticism, much of it arguing that he had paid too little attention to the role of coercion in subduing the working classes and pointing to evidence of shop-floor struggles which revealed this 'consensus' to be an illusion. For a brief coverage of this debate in English see F. Andreucci, 'Subversiveness and Anti-Fascism in Italy', in R. Samuel (ed.), *People's History and Socialist Theory* (London, 1981). A complex and subtle contribution to this hitherto bi-polar and heated debate emerged in the work of Luisa Passerini. Using oral testimonies as her main source material she has argued that varying degrees of consensus and resistance could coexist in the same person, often in a seemingly contradictory way. Some of Passerini's writings are available in English. See e.g. L. Passerini, 'Work Ideology and Consensus under Italian Fascism', *History Workshop*, 8 (1979); Passerini, *Fascism and Popular Memory* (London, 1987).

It was the voice of the people itself which was saying to him 'I'm here near you, I can hear you, I can understand you, I love you, and I want you to understand how much I love you.'[15]

But such romanticized reports describing the endless enthusiasm of the masses for Mussolini were, of course, commonplace in inter-war Italy. Many of them were largely imaginative journalism, and this welcome for the Duce may have been little more than stage-managed pageantry. A different light is shed on this incident by a report in the antifascist newspaper *La Libertà* which described the careful preparations made for Mussolini's visit. Police were placed throughout the building and all the work-force searched. As an additional precaution eighty workers were kept away from the others for the duration of the ceremony and confined to a separate room.[16]

It is clear that not all the work-force were as dedicated to Mussolini as the management propaganda maintained. A Fascist presence did persist in this firm throughout the twenty years of Mussolini's rule, but what little evidence has survived does not suggest that the proportion of activists here was particularly high except in the early 1920s when the firm was very small. In this period, two of the founding members of Sesto's *fascio* (quality control inspector Vasco Carlini, who was head of Sesto's Fascist squad, and Luigi Delor, a clerk) were employed here, and in the firm itself they formed a *Gruppo Aziendale* (Factory Group) in 1921, the first of its kind in Sesto. This was reported in the Fascist paper *Il Popolo d'Italia*:

Yesterday a sizeable core of Fascists, composed of clerks, foremen, and workers from Fabbrica Italiana Magneti Marelli, met to found a Fascist group inside the factory itself, composed of people who are all regular members of the various *fasci*. Its main aim will be Fascist propaganda inside the firm.[17]

Another report later the same year shows that the group's activities met with some success, as they managed to get one of

[15] 'Il Duce fra noi', *SB* 6 (1930), 3–4.

[16] 'La visita agli operai alla Marelli', *La Libertà* (newspaper of the Concentrazione antifascista di Parigi) (7 June 1930). Cited in Longaretti, 'Il fascio di Sesto San Giovanni'.

[17] 'Il fascismo negli stabilimenti sestesi', *Il Popolo d'Italia* (1 June 1921).

their number, a Fascist activist named Rabino, elected to the firm's Internal Commission.[18]

However, this early strength does not appear to have been built upon, and in subsequent years the Magneti Fascist group never grew particularly big. Oral testimony and pictures in the factory newspaper suggest that this was just a smallish group of about thirty clerks and workers, mostly male. By the time of the Liberation there were about fifty *repubblichini* scattered throughout the various shops and offices of the firm,[19] and in 1945 the number of people purged from the A Plant was fifty-six, thirty-eight of whom lost their jobs because they were 'persons who had been involved in Fascist and collaborationist activity', whereas the others were simply 'fired without notice because they are considered undesirable and harmful to the proper functioning of the firm'.[20]

It is highly probable that the number of party members was greater than this, although exactly how many there were is far from easy to determine. The evidence from the interviews in this respect is confusing. Some interviewees thought that it was not a prerequisite for employment, whereas others simply said they could not remember. Some, however, asserted that the party card was definitely a help in getting work and in certain cases a condition of employment, although it seems that this was due less to the firm than to pressure from the unions who controlled the local labour exchange. According to one testimony, 90 per cent of the workers in the firm possessed a PNF card, but this was not because they believed in Fascism: 'It wasn't that

Support ? Party Card

[18] 'Operai fascisti nelle Officine Marelli', *Il Popolo d'Italia* (19 Nov. 1921).

[19] 'At the FIMM there was a lot of spying done by the notorious Battaglia and his cronies. There were a lot of *republicchini* at the FIMM, about 50 of them and all of them petty hierarchs and 100% supporters of the Germans. In every single office or shop there was one of these miserable characters and you can imagine how difficult it made our work' (Valerio, 'Vita ed attività clandestina del CLN aziendale FIMM', ACLNL, b. 4, f. 13, p. 6).

[20] 'Comunicazioni 06 e 07 del CLNA dello stab. A, 9.5.1945.', ACLNL, b. 4, f. 13. In the personnel archive 3.6% of the *fogli d'assunzione* in the personnel files were marked 'PNF yes' and 0.6% were marked 'PNF no'. 95.8% of the files were not marked at all in this manner. This information is of dubious value, as it is unclear what criteria were used for noting whether or not someone was a Fascist Party member. Maybe this simply depended on which particular clerk filled in the form. It may mean, on the other hand, that only real activists were noted. (Source: APMM, figures compiled by myself.)

many of them were Fascists. They joined the *fascio* to get work. Because in those days they called it the meal ticket' (Santino). According to another testimony Fascist membership was compulsory, but after Quintavalle lost his own party card the firm did not bother so much and 'Magneti Marelli became more of an open door in those years' (Costantino).

One persistent feature of all the oral testimonies was that, whether they remembered the card as being compulsory or not, personally all those interviewed were reluctant to admit that they themselves had ever had one. This underlines the fundamental problems with using oral sources to investigate such issues. Few present-day residents of Sesto San Giovanni can be proud of having had a Fascist party card, even against their will. Of those I interviewed only two admitted to this, both of whom stressed that they had received cards automatically as they progressed up through the different youth organizations at school like many other Italian children.

FASCIST UNIONS

For most of the duration of Mussolini's regime, the Fascist 'trade unions' were caught in the uncomfortable position of trying to recruit those whose interests they could do little to protect. Their very contradictory and difficult position *vis-à-vis* the employers was well summed up during the 1925 metalworkers strike by Nino Cuzzeri, Milanese provincial secretary of the Fascist engineering union, who stated during a meeting that 'after three years of patience we have at last come to realize that the co-operation with the industrialists that we have so sincerely hoped for, is impossible.'[21]

After this the unions' position deteriorated further and by the late 1920s their aspirations to be an effective voice in the regime was totally defeated. Although the Pact of Palazzo Vidoni (2 October 1925) gave them their much-sought-after monopoly to represent the workers, it simultaneously robbed them of their victory by abolishing the Internal Factory Commissions. Following this they also failed in their attempt to obtain legal

[21] Cited in A. De Bernardi, 'Operai, sindacati e regime negli anni venti: il caso di Milano', *Società e storia*, 40 (1988), 345.

recognition for their individual representatives in factories (*fiduciari*). With the *sbloccamento* (unblocking) of 1929 the unions' central body was broken down into six separate organizations, which severely reduced their ability to put pressure on the government. After this series of defeats the unions became largely bureaucratic organizations with only the long-winded ineffectual Labour Court to defend their members.[22]

The few fragmentary pieces of information which have survived on the role of the unions at Magneti Marelli, suggest that there, as in other firms, they were cold-shouldered by the management. Like many other industrialists who supported the regime, Benni and Quintavalle's vision of a Fascist Italy did not embrace the Fascist unions. Enthusiastic about including endless Fascist rhetoric in the activities of the firm, they were less keen to allow a group of employees to challenge their authority on the shop-floor, and Benni himself, in his role as head of the *Confindustria*, proved a determined opponent of a union presence inside actual firms.[23] The discontent of the firm's union activists is evident in the following passage from a letter dated January 1929:

> we have to complain about the systematically hostile attitude of the Magneti Marelli management towards all the union members, which is shown openly in threats of sacking, dismissals from certain jobs, accusations of failing to support the interests of the firm.[24]

Another similar report later that year tells the same story. This report by Nino Cuzzeri (by now Union Inspector) tells of the

[22] On the Fascist unions see C. Schwarzenberg, *Il sindacalismo fascista* (Milan, 1973); G. Salvemini, *Under the Axe of Fascism* (London, 1936); G. Sapelli, 'Per la storia del sindacalismo fascista: tra controllo sociale e conflitto di classe', in *Studi storici* (1978); De Bernardi, 'Operai, sindacati e regime'. On the question of the 'fiduciari' see A. Aquarone, 'La politica sindacale del fascismo', and on the Labour Court see G. C. Jocteau, 'La magistratura del lavoro nello stato fascista: le contraversie collettive (1926–35)', both in A. Aquarone (a cura), *Il regime fascista* (Bologna, 1974).

[23] A. Lyttleton, *The Seizure of Power: Fascism in Italy 1919–1929* (London, 1973), 326.

[24] Letter from Sindacato Provinciale Impiegati e Tecnici delle Industrie Meccaniche to the Commissario Straordinario dell'Ufficio Provinciale CNSF, ASM, Gab. Pref., cart. 43, b. 1074, 'Sindacato fascista', f. 'Situazione sindacale provincia metallurgia', sf. 'Sesto San Giovanni', 25 Jan. 1929.

problems faced by the Magneti *Gruppo Aziendale* after it had been placed under the jurisdiction of the Fascist unions:

> As far as the *Gruppo Aziendale* (whose delegate is Sig. Marchesi) is concerned, it seems that no one even notices its existence. But what can you expect, we are, after all in Marelli's firm (in other words Benni's).[25]

This was serious for them since, with the exclusion of their own representatives (*fiduciari*) from the shop-floor, the unions had to make do with a presence via the *Gruppo*. For ten years the problems faced by the unions in this firm appear to have continued, according to a report by a Magneti Communist which appeared in *Lo Stato Operaio* in 1939:

> The factory management doesn't want the workers to go to the Fascist unions when they have a problem. In a notice pinned up in the shops, the management announced to the work-force that if they had any queries about wages, disciplinary matters, etc. they should go directly and personally to the Personnel Office and not to the unions.[26]

This suggests that the unions found it far from easy to operate in this firm. This does not mean, however, that they were entirely powerless. Indeed the extract just quoted suggests that the management were in fact worried about that influence.[27] Oral testimony does indicate that the unions were sometimes willing to assert themselves *vis-à-vis* the management, albeit generally only on behalf of clearly Fascist workers. Many of the work-force failed to fit into this category.

[25] N. Cuzzeri, 'Situazione generale operai metallurgici, 12 ottobre 1929', ASM, Gab. Pref., cart. 1061. Being concerned with propaganda rather than economic issues like the unions, the *Gruppi Aziendali* were initially encouraged by the employers: 'It is well known how the industrialists have always accepted and therefore recognized the existence of the representatives of the *Gruppo*. In almost every firm where there is a *Gruppo*, the management has given them both moral and financial support.' (Unsigned memorandum in ASM, Gab. Pref., cart. 1061, probably written by Cuzzeri, 1929.)

[26] 'Dati e considerazioni sulle condizioni della classe operaia italiana: le condizioni di lavoro degli operai della "Magneti Marelli" ', *Lo Stato Operaio* (15 June 1939), 245.

[27] This is backed up by a memorandum found in one of the personnel files dated 1942, in which the writer (a clerk or manager) mentions that a worker needed to be given a higher grade as his work had changed. One of the stated reasons for this regrading was 'to avoid possible trouble from the unions'. (Promemoria Ufficio manodopera, 24. 06. 1942, APMM.)

ANTIFASCISTS

Police and other records yield little information on whether the Magneti workers were involved in the labour militancy and the struggle against the rise of Fascism in the early 1920s, probably due to the small size of the firm in these early years. One incident that has been recorded was a strike in March 1923 which broke out over a number of issues, including management attempts to sack numerous workers and rehire them only after they had given an oath of loyalty to the firm.[28] After this strike political activity was gradually forced underground. A series of reports in *Unità* in the 1920s shows that even after Fascism took power a Communist presence persisted. This continued sporadically in the 1930s. Magneti worker Carlo Seveso, for example, was incarcerated for four years by the Special Tribunal (the Fascist court for political crimes) in 1931. Communist Party clandestine factory groups were mentioned in 1930, 1931, 1932, and 1934 in documents kept in the Communist Party Archive,[29] and from 1934 there was a link with the 'Gruppo Cellule-Nord Milano' and with the five others in the town based 'Vittorio' group led by Darles Fusetti. By 1937 there were three workers' groups and one of clerks.

Sheer power of firm. management

Not overt outlets for opp.

These groups, as in other factories, were divided into two sections. One, the 'legal' section, concentrated on attempting to infiltrate the Fascist unions and Dopolavoro, and encouraging other workers to take industrial action. 'Legal groups' played a key role in organizing the stoppages mentioned in Chapter 4. There was also a more secretive group which focused on disseminating antifascist propaganda, particularly aimed at the young. This section had three members only one of whom was in contact with other groups for security reasons. According to one account, in the 1930s antifascist activism was strongest in shops such as the Professional Radio Shop where there were

[28] The strike was also over suspensions, finings, wage cuts, extended work hours, and recognition for the internal commission. See Bell, *Sesto San Giovanni*, 168.

[29] 'Notes sur des conversations avec des délégués de base', IV Congrès PCI, APCI, f. 933, fo. 78. For 1931 see APCI, f. 933/2. For 1932 see APCI, f. 1061/2. For 1932, the information is more detailed—there were two party members both from the same shop and three 'trade union members'.

many skilled male workers.[30] They did not, however, have a total monopoly. In 1934 the Communist Party was in contact with a female clerical worker.[31]

Reflecting the general growth of the Communist Party throughout Italy, in the late 1930s and the early 1940s, clandestine activity increased with an expanded Communist presence. According to one party activist, Antonio Mantovani, by 1939 'at Magneti Marelli there were about forty of us Communists in factory groups, without counting the sympathizers who weren't actually members. We were also linked with other neighbourhood groups.'[32]

In the 1930s, therefore, there was some almost continuous antifascist activity, although it was still small. The time when this movement became really significant was with the advent of the Second World War, the period of the Resistance. The history of the Resistance has generated an enormous amount of historical writing and debate, and it is clearly beyond the scope of this book to enter into a detailed examination of such a complex period. But Magneti was located in 'Red Sesto', and some discussion of this is essential to an understanding of the history of this firm under Fascism.

THE RESISTANCE

Magneti Marelli had its part to play in the momentous events of the Resistance. The Communist strength of the late 1930s had been momentarily disrupted when 'Elli', a Communist working in the Radio Shop, talked when arrested and many of the Communists in the firm were arrested, or, like Antonio Mantovani, forced to flee to political exile in France.[33] But by late

[30] C. Picolotti, undated 'Memoria', PCI, Sez. 'Lenin' Magneti Marelli SSG. This document was shown to me by the author.

[31] This is the clerk who sent the secret reports quoted in Chapter 4. 'Costantino', in a clandestine report, refers to 'a woman who is a good sympathizer working as a shop secretary in the Mag. Mar. She is the ex-girlfriend of the comrade that I saw previously': 'Da Costantino' F.G.C.I., 14 Jan. 1934, APCI, f. 1241, fo. 44.

[32] 'L'Antonio Mantovani', in F. Alasia, *La vita di prima* (Milan, 1984), 160.

[33] Elli was quite important, as he was the leader of the Communist group 'Gruppo Nord Milano'.

1941 clandestine activity had begun to recover and, as in other large firms in the area, the numbers involved grew steadily. After 25 July 1943 when Mussolini fell, the Magneti work-force elected its own internal factory commission which began immediately to purge some of the most overt Fascists, including Bruno Antonio Quintavalle who was replaced by his brother Umberto. After 8 September the commission was forced underground by the arrival of the Nazi troops and a factory-based CLN (National Liberation Committee) was set up with the help of Umberto Fogagnolo, one of the 'Loreto martyrs'.[34]

In the final months of the war the armed resistance grew in importance. Some left the factory to become partisans, either in the mountains or in urban areas. But in the firm too preparations for the insurrectional strike that would end the war began with the creation of some SAPs (*Squadre d'azione patriottica*). These were initially created to defend workers during strikes, but later took a more offensive role.[35] At first Magneti workers, together with some from Ercole Marelli, were involved in the 109th SAP Garibaldi Brigade, but the summer of 1944 saw the creation of the A plant's own brigade, the 129th Garibaldi SAP Brigade 'Carlo Mandel' commanded by Aldo Pastrello (code-name 'Venezia') and with Emilio d'Agaro as political commissioner. The 109th and the 129th worked closely with each other and with partisans in the mountains, to whom they sent supplies and equipment including radio transmitters. At the N plant

[34] The Loreto martyrs were a group of antifascists, two of them employed at Ercole Marelli, shot by the Fascists in Piazzale Loreto in Milan on 10 Aug. 1944. The shooting had a great impact on public opinion, as most executions had been in private before then.

[35] The SAPs had a varied role. Most of the members were involved in actions such as the distribution of newspapers and propaganda, writing antifascist graffiti, sabotage, and the prevention of arrests after strikes. Some took part in more military activities such as the killing of spies discovered in factories. Unlike the exclusively communist GAPs, the SAPs were open to all antifascists and even in big factories the majority of the members were non-communists. They constituted the main way in which factory workers got involved in the Resistance, since many of the mountain partisans were deserters and workers were less likely to have been called up in the first place. Although staying at home and joining a SAP was in some ways easier than mountain guerrilla warfare, political work in factories was more isolated and lacked the camaraderie of the mountains. On the SAPs in Sesto and Milan see L. Borgomaneri, *Due inverni, un'estate e la rossa primavera: le brigate Garibaldi a Milano e provincia (1943–1945)* (Milan, 1985).

many workers joined the 110th Garibaldi SAP Brigade 'Beppe Ottolenghi'.

Undoubtedly the Communists were the most important antifascist grouping within the factory. By April 1944 they had 53 party members at the A plant (four of them women) and 23 (three of them women) at the N.[36] But other groups did emerge too. The Action Party set up their own group, and some Socialists, linked to Umberto Fogagnolo, were co-operating with the Communists by 1944 in the N plant.[37] The Socialists had their own partisan section at Magneti, a unit of the 49th Matteotti Brigade 'De Zorzi'. There were also a number of Catholic activists, although they do not seem to have organized separately.

The activities of these groups, Communist or otherwise, were similar to those in many other firms, a pattern already amply documented in the huge numbers of books and articles which tell the complex story of the Italian Resistance. In the early years of the war they concentrated on the manufacture of and distribution of leaflets calling for peace and an end to Fascism. Already by autumn 1942 small strikes were occurring. Prominent in these were women from the Shearing and Radio Shops who demanded higher food and fuel rations.[38] But the struggle intensified enormously once the tide began to turn in the war with the defeat of the Germans at Stalingrad. In this firm, like many others, the clandestine groupings organized strikes and distributed antifascist literature. Many were also involved in sabotage, making components which burnt out after only a few hours use or delaying the construction and delivery of products needed for military purposes. One example was the extensive delay in the manufacture of a large radio transmitting station for the German army.[39]

Although the smallest of the large factories of Sesto (the others being Breda, Falck, Pirelli, and Ercole Marelli), virtually all the Magneti work-force (including clerks) participated in the great strikes of 1943 and 1944, often, as in the huge strike which

[36] See Pontecorvo, 'Rapporto sulla situazione', in Scalpelli, *Scioperi e guerriglia*, 97 and 101.

[37] Ibid. 101. By Dec. 1944 the Magneti Socialist group had 12 members (half clerks and half workers) and 50 sympathizers (Petrillo and Vignati, *Città e fabbrica nella Resistanza*, 72).

[38] Picolotti, 'Memoria'.

[39] Valerio, 'Vita ed attività clandestina', 7.

began on 13 December 1943, being among the first firms to down tools. These strikes were precipitated by the fast worsening economic conditions, but were at the same time political and antifascist in nature.[40] They were, in this firm as elsewhere, to a great extent the product of exhaustion with the war and its food and fuel shortages and anger at the German occupation.[41] In some cases the strikers' demands for food or fuel were met. The strike of December 1943, for example, proved particularly successful and Magneti managed to get more concessions from the Germans than other firms. Other strikes met with a tough attitude, as in November 1944 when a strike at the N Plant was countered by a lock-out.

April 1945 saw the Liberation of the North of Italy by the Allied troops, preceded by mass uprising—the 'Insurrection'. As the Germans left for Switzerland the SAPs occupied the Sesto factories to defend machinery from dismantling or even destruction. At Magneti they were poorly armed, but this proved unimportant since by this time the Germans were on the run and, although Sesto did see some fighting (for example at the Pirelli factory), there were no huge pitched battles like those which preceded the Liberation in some other parts of Italy. Here the 'Insurrection' was over in three days.[42] On 28 April, the day after the fighting ended, the Allied troops arrived.

With the departure of the Germans the time had come to settle old scores. Deciding who was to be punished was not necessarily an easy task given the rapid changing of sides of many in the final few months of the war. The directors fled, although the foremen generally remained, which suggests that

[40] A huge literature has been written about these strikes. See e.g. P. Spriano, *Storia del Partito Comunista Italiano*, v: *La Resistenza: Togliatti e il partito nuovo* (Turin, 1976). On the strikes in Milan see E. Gencarelli, 'Partito e classe a Milano negli scioperi del 1942–44', *Italia contemporanea*, 114 (1974); Scalpelli, *Scioperi e guerriglia;* L. Ganapini, 'Milano, autunno 1944: la svolta delle lotte operaie', *Italia contemporanea* (1975); Ganapini, 'Lotte operaie Milano', in G. Quazza (ed.), *Operai e contadini nella crisi italiana del 1943–44* (Milan, 1974); Ganapini, *Una città, la guerra: Milano 1939–1951*) (Milan, 1988); G. Brambilla, 'Il movimento operaio nel 1943 a Milano', *Critica marxista* (1974). For a detailed account of the events of the 1943–5 period at Magneti written by a member of the antifascist CLN, see Valerio, 'Vita ed attività clandestina'.

[41] Such shortages dogged Milan throughout the war and a black market emerged as early as 1941. See Ganapini, *Una città, la guerra*, 26.

[42] On the 'Insurrection' in Sesto see the eyewitness account by B. Mauri reproduced in Borgomaneri, *Due inverni, un'estate*, 259–63.

they believed that they had little to fear. The personnel manager and the head of the Pay Office (Guagliumi and Marchesi, who were both noted Fascists) were killed by partisans away from the factory. They were not on the premises since they had been sacked in 1943. Partisans also killed another man, a clerk who had helped in the rounding up of Resistance activists.

Michele Soliveri, who was by then head of the N plant, was the only one actually to die in the factory itself. A number of the interviewees asserted that he was not necessarily the most hated of the management but simply unlucky for being around at the time. Unlike the Quintavalles, who probably would have met the same fate had they stayed, he decided not to run. Some of those I interviewed believed that he should not have been killed, since part of the anger against him stemmed from purely personal grudges. But Soliveri may not have been as innocent as these testimonies suggest, and the killers certainly believed their anger justified. The accusations against him were serious. Soliveri was singled out because he was the person who had actually signed the list of people to be taken away to work in Germany. Some of these had suffered terribly and returned as half-starved skeletons. Others failed to return at all. He defended himself by claiming that on the contrary he had actually managed to reduce the numbers from 300 to 160, and that he had tried to include only people without dependants. The internal commission of the factory decided on 27 April 1945 that Soliveri could go free providing he left the factory. He was cleared partly because the CLN of Caravaggio, his own village, came to testify on his behalf. However, he did not go to Caravaggio but stayed overnight in his lodgings in the factory. This proved his downfall, as not all the Resistance activists accepted the decision of the commission:

> But they were angry because they said that he had certain defects as regards women or something like that. But it was something minor, not something to kill him for . . . With the women clerks in the factory. (Giuseppe)

In the morning some young partisans dragged him from his lodgings and killed him.[43]

[43] Later the four young men responsible were arrested, tried, and given long prison sentences. Eventually, however, they appealed successfully after serving some of this time, and their action was recognized as an 'act of war'.

As must be clear from this account, like the other factories of Sesto, Magneti witnessed a dramatic series of events in this period and some of its work-force risked their lives in the antifascist cause. Yet in certain respects the experience of Magneti in the Resistance was dissimilar to that of some of the other Sesto factories. The figures on the number of workers deported to Germany for political reasons contrast greatly. At the Breda heavy engineering firm about 350 were deported. The number from the Falck steelworks was 105 and from the Pirelli rubberworks it was 243, compared with only three from Magneti.[44] The heroic but tragic list of the numbers of those who died in the fight against Fascism shows a similar picture: 317 Sesto people (all men) gave up their lives. Some were political deportees who died in concentration camps. Others died fighting, or were shot, or died in prison. Of these, nearly half (134) came from the Breda plant, 69 were employed by Falck, and 41 by Pirelli, whereas only two—Carlo Beretta and Carlo Mandel—were from Magneti.[45]

Obviously the smaller size of Magneti compared with giants such as Breda (12,000 workers by 1944) and Pirelli partially explains this situation, but a number of other arguments merit consideration to understand why the Resistance record was rather different here from that in other local firms.

One explanation might be that it was less the level of resistance that was different than the level of repression. It is possible that the Germans and Fascists were more light-handed in their treatment of 'unrest' in this firm because they were reluctant to clamp down too much on a female-dominated work-force. A report in early 1945 written by Zanuso, the commander of the garrison of the Republican National Guard (GNR) of Monza, suggests that it was indeed men who were seen as the real problem:

[44] *Sesto San Giovanni nella Resistenza*, p. 20 n. The sending of workers to Germany for which Soliveri was killed was for forced labour and not directly for political reasons. The figures for Magneti's 'parent' factory, Ercole Marelli, were also low; 11 were deported. This, it is said locally, was due to the fact that one of the engineers successfully protected the workers. (I am indebted to Giuseppe Vignati for this information.)

[45] These figures are compiled from lists in Vignati, *I ribelli al governo della città*, 71–5. Of the remainder of the 317 dead, the majority were not factory workers although some came from other, mostly much smaller, factories.

this totally subversive industrial centre is an absolute curse! It is the cancer of Lombardy. This red town should be completely destroyed . . . The male population should be transported to Germany, leaving behind only women, old people, and children.[46]

Furthermore, one oral testimony (Ercole) asserted that deportations were few simply because the work of Magneti was so important to the German war effort. He pointed out that the strikes were as well followed here as at Breda, but the Breda steelworks mattered less to the Germans because the Ruhr was still functioning. This may well have been a significant factor, but no other evidence has come to light to prove such a theory.

Whether the level of repression meted out was more lenient, or not it may be that there was quite simply less to repress. In other words the Magneti workers may have been more afraid, more conservative, or even more pro-Fascist than those in other firms. Previous chapters have shown how the firm managed to create a sense of loyalty amongst a section of the work-force, and one clandestine document does refer to those in the N plant as being 'cold and reactionary'.[47] However, this is an isolated comment and certainly not typical of descriptions of the labour force in this factory. As was argued above, the extent to which the recruitment of at least nominal PNF members was favoured is uncertain, and the political views of the majority of the work-force can be only guessed at given the scarcity of evidence.

What does seem clear, however, is that the particular composition of the labour force in this firm did have an impact on the pattern of political activism. Giuseppe Vignati has argued that one crucial factor may be the influence of large numbers of rural-born workers. He asserts that in certain Sesto factories, including Magneti, such workers may have diluted the Resistance movement:

[46] Ten. col. G. Zanuso, 'Relazione di zona', 21 Feb. 1945, Archivio privato Lazzarini, AISRMO. Cited in Vignati, *I ribelli al governo della città*, 46.

[47] 'Today we organized a "flying meeting" at Magneti Marelli. Our members from the SAP took part. It was very successful, even though the speaker wasn't exactly brilliant. The workers from the 4th plant talked about it enthusiastically for a long time afterwards, even though they are a most cold and reactionary group of workers.' Rapporto interno PCI, photocopy in AISRMO, Res. b. 2 (fondo Martinini); reproduced in Petrillo and Vignati, *Città e fabbrica nella resistanza*, 70.

> the migrant workers who came mostly from areas like Ferrara were people who had probably managed to get work in the factories by having a reference from the local Fascist Party branch and they tended, therefore, to weaken the antifascism of these work-forces.[48]

Vignati does qualify this as simply a hypothesis and, in fact, as far as Magneti is concerned, it is hard to tell whether it was significant. As was shown in Chapter 5, the origin of migrant workers was varied and the data are far from conclusive. Furthermore, although there were workers who tried to make use of recommendations in attempting to get work in the firm, most of these were local applicants, and many of the references filed in the Personnel Archive were not written by Fascists. Vignati's hypothesis, therefore, can be neither proved nor disproved. It may, however, have been a relevant factor with regard to a small section of the work-force.

Perhaps more significant were the differences between the skill, age, and gender composition of the work-forces of the different Sesto factories. In firms like Breda, where the work-force was predominantly male and skilled, political activity was easier since such workers were in short supply and to sack the activists might have endangered the firm's productive capacity.[49] If these workers did lose their jobs then they could more easily find new ones than the unskilled or semi-skilled. So Breda workers were more likely to be able to afford to take risks. At Magneti, on the other hand, a much larger proportion of the work-force were unskilled or semi-skilled and highly vulnerable since there was a sizeable pool of labour ready to replace anyone who stepped out of line. Oral testimony shows that many of those who did get involved at Magneti were indeed skilled workers. Ercole, for example, asserted that the Toolroom (which

[48] Vignati, 'Un'opposizione dura a morire', in Petrillo, *La città delle fabbriche*, 101.

[49] The power of the skilled Breda workers to negotiate is well illustrated by an incident in 1934 when, in order to get the bonus rates raised, a number of workers individually gave in their notice, each with a different explanation. The manager concerned realized what was going on and asked them all to come to see him. Once he had ascertained the reason for their protest he gave in to their demands. On another occasion a delegation of three workers simply went to the management headquarters and put forward their demands. They succeeded in obtaining a pay rise. (Oral testimony of Carlo Ciceri cited in F. Alasia, *Gaetano Invernizzi dirigente operaio* (Milan, 1976), 59–60.)

employed only highly skilled male workers) was the most 'red' shop in the factory.

One very important thing is that in the Toolroom about seventy per cent were red, well antifascist more than red. But its always been the fitters who were the antifascists. . . . I think it was due to one important factor, fitters had a certain amount of education . . . they had to know about design, mathematics . . . That meant that the fitter was more emancipated, he saw things more objectively, and that's why he was an antifascist. In my shop . . . the top worker was a Fascist, but very few of the others were. When 1943 came, it was really strong there, all these antifascists appeared . . . in the Toolroom there was a big underground support network. They used to meet to discuss things, about the Insurrection, to organize the strikes.[50] (Ercole)

Other shops where there was a great deal of resistance activity (notably the Brake-making Shop and the Radio Receiver Shop) had more mixed work-forces and the skill composition of activists was more varied. None the less, even here the most skilled seem to have predominated in antifascist activity.

The typical Magneti worker can also be contrasted in age with the Breda workers, many of whom were older and some of whom could still remember the pre-Fascist period and the workers' struggles, whereas fewer of those in the mostly youthful work-force at Magneti had personal political experience of this kind. Many of these workers had grown up under Fascism and had known no other government than Mussolini's. As such it must have seemed immovable, not to be toppled nor opposed, but simply a reality of life. Excellent!

Another undoubtedly important factor was the gender composition of the work-force. Maybe it was simply the presence of large numbers of women which led to a 'watering down' of the Resistance movement in this firm. It might be argued that women were quite simply more conservative than men.[51]

[50] According to Borgomaneri in many factories the toolmakers headed the Resistance. See Borgomaneri, *Due inverni, un'estate*, 177.

[51] Maria-Antonietta Macciocchi in her book *La donna nera: consenso femminile e fascismo* (Milan, 1976) has gone further and argued that many women were pro-Fascist. Macciocchi's book is an important one as it touches on a vital issue and does include some interesting information, but its arguments are, at best, problematic. She tends to make too many broad and unproven generalizations and over-simplistically equates passive acceptance with support. Even more dubiously, her explanation of this consensus is that it is a result of women's masochism. Whilst psychological approaches to history can be illuminating,

Indeed, many left-wing male activists held this view themselves at the time, fearing the influence of the Church on women's lives. However, to simply brand women as a conservative force is to simplify their experience of this period. Women did contribute to the Resistance, but male and female roles differed greatly. To evaluate by figures such as the numbers killed or deported misses the essence of women's own specific contribution to the struggle. It might be better to argue that the male-dominated forms of resistance activity were less at Magneti than at Breda. In terms of the specifically female contribution to the resistance, on the other hand, the Magneti women were a far from insignificant force.

WOMEN AND ANTIFASCIST POLITICS

Long before the beginning of the armed Resistance, women in the Milanese area frequently took an active part in, or even themselves led, strikes and agitations, both in and out of the workplace.[52] Amongst factory workers there were many examples of militant action. The textile industry, a huge employer of women, was the scene for a great deal of industrial strife. Perhaps the most dramatic example was the violent series of strikes by female textile workers in Legnano faced with the problems of an industry in serious crisis. In one incident, provoked by threatened wage cuts and the sacking of fifty women, 500 women stopped work, sang the 'Red Flag', and drove the police away by stoning them. Practically the entire police force of Milan was needed to restore order.[53]

During the Second World War in this area, as in other parts of

Macciocchi's use of them is at best unhelpful. In relying on these ideas she neglects the complex social, cultural, and historical reasons which need to be considered to understand how women could support (or apparently support) oppressive misogynist regimes.

[52] e.g. to cite but one incident from many, on 17 Jan. 1933 in Brusuglio (near Milan) there was a demonstration which '. . . involved all the women from the surrounding villages. The women elected a committee who went to see the mayor and in front of him they tore up their tax forms in a protest against taxes. The mayor and the prefectorial commissioner managed to calm them down by promising to cut taxes by 50%' (Bay, 'Rapporto Milano città, settore Brusuglio', APCI, f. 1061, fo. 33).

[53] 'Dal nostro funzionario—Milano 10 aprile 1932' (unsigned), APCI, f. 1061/2.

Italy, there were many highly courageous women who contributed to the Resistance movement against Fascism. Although accurate records do not exist, it is possible that the number of women involved was equal to the men. Furthermore, unlike many of the men who got involved because they were faced with a choice between being called up to fight a war they no longer believed in and the life of a partisan, the women were all real volunteers. They could have chosen to stay at home. Women from many different walks of life participated in a variety of roles, they were present in the work of every political grouping within the Resistance, and their reasons for involvement were many, ranging from a wish to hasten the end of the war to a desire, among some of the younger women, for excitement and a wish to break out of traditional gender roles.[54]

In the Sesto Resistance movement women played an important part. In June 1942 a large women's demonstration outside the Sesto Town Hall for the distribution of potatoes started the wave of protests against the war and Fascism.[55] In the factories too, numerous women took part in the strikes of this period. At times they took the lead. In the strikes of September 1944, according to a clandestine Communist document, 'Once again in these strikes the women have participated fully; in certain factories they have led the way and convinced the men to stop work.'[56]

In the Breda factory it was women who more than once 'gave the notorious director Bovone, the collaborationist informer who was terrorizing the factory, a good kicking with their clogs'.[57] In one incident in April 1944 a group of Sesto women, with the help of the railway workers from the goods depot,

[54] On women's role in the Resistance see e.g. 'Donne (nella Resistanza)' in A. Boldrini, *Enciclopedia della Resistenza* (Bologna, 1980); B. Guidetti Serra, *Compagne* (Turin, 1977); M. Alloisio and M. Ajo, *La donna nel Socialismo Italiano tra cronaca e storia (1892–1978)* (Bari, 1978); C. Ravera, *La donna italiana dal primo al secondo Risorgimento* (Rome, 1951); B. Taddei, *Donne processate dal Tribunale speciale 1927–43* (Verona, 1969); M. Alloisio and G. Beltrami, *Volontarie della libertà* (Milan, 1981); A. M. Bruzzone and R. Farina, *La Resistenza taciuta: dodici vite di partigiane piemontesi* (Milan, 1976); G. Casadio and J. Fenati (eds.), *Le donne ravennati nell'antifascismo e nella Resistenza* (Ravenna, 1977).

[55] Ganapini, *Una città, la guerra*, 34.

[56] Federazione milanese del PCI, 'Relazione interna', 08.09.1944, AISRMO, Res. b. 1 (fondo Fontanella), p. 9. Partially reproduced in Petrillo and Vignati, *Città e fabbrica nella resistenza*, 57.

[57] *Sesto San Giovanni nella Resistenza*, 20.

managed, using burglary tools, to open three sealed carriages of a temporarily stopped train and set free more than a hundred youths who had been rounded up by the Nazifascists during the battle of Benedicta (Alessandria) and who were on their way to concentration camps.[58]

There were also a handful of women who took up arms in the Garibaldi Brigades. Most women, however, were involved, as in other parts of Italy, in a more auxiliary role. Few bore arms, but many carried messages and antifascist literature in bicycle tyres and handbags, acted as spies, typed messages and news-sheets, sheltered partisans, collected money and clothes for the partisans, and organized a hospital in the final period of the Insurrection.[59]

WOMEN AND THE RESISTANCE AT MAGNETI

At Magneti too, some of the women played a part in the development of antifascist activities, and many of them were far from timid during the agitations and strikes of the final years of Fascism. This is evident in a Communist Party report dated December 1944 describing the unrest in this firm:

> the discontent after the distribution of the Christmas bonus is increased by the comparison with the better treatment of the Ercole Marelli workers. The Agitation Committee immediately sent a large delegation to the top management, but they were not allowed in. It has been suggested to the Agitation Committee that they put up a notice and do not give in. . . . *The women are leading the protest.*[60] (my italics)

Apart from such collective actions, certain women also made more individual contributions to the Resistance. There were, for example, some who acted as messengers, such as Giovanna Camisa, who carried vital messages between the CLNs of Ercole Marelli and Magneti. Another woman, referred to simply as 'Chiara', was remembered by one of the interviewees (Bruna) as making an armed arrest of one of the directors during the Insurrection. Women also did some of the dangerous work of

[58] Vignati, 'Prologo: "Il cancro della Lombardia" ', in *I ribelli*, 53.

[59] *Sesto San Giovanni nella Resistenza*, 20–1.

[60] Rapporto interno PCI, photocopy in AISRMO, Res., b. 2 (fondo Martinini). Reproduced in Petrillo and Vignati, *Città e fabbrica nella Resistenza*, 71.

spying, work for which they were particularly well suited due to the relative lack of suspicion they aroused from the Germans and Fascists. One of the interviewees recalled how one of her fellow workers had a love affair with a Fascist in order to collect information which she then passed on to the partisans. Another example was a female doorkeeper who used to send messages to alert people if they were in trouble or about to be called up as forced labour by the Germans.

There were also a handful of remarkable female activists, mostly little more than girls. One of these was Isa De Ponti (clandestine names 'La Narva' and 'Barbara'), who worked at the N plant in Crescenzago. Born into a left-wing family, she was already working for the antifascists in 1936 at the age of sixteen, writing letters in invisible ink, typing, carrying messages on her bicycle, and distributing clandestine literature. She joined a Communist group which was centred on the Breda plant. At this young age she had already met Vera Ciceri, herself an experienced Resistance worker.[61] Some years later, after spending a few years in prison, Vera Ciceri returned to Sesto with directives from the Communist Party to organize the women of the area. By this time Isa had left the job she had in a spinning mill and was at Magneti N. Following Vera's instructions she helped found a *Gruppo di difesa della donna e per l'assistenza ai combattenti della libertà* (GDD—Women's Defence Group in Support of the Freedom Fighters) at Magneti.

Soon after Isa had to give up this work and flee the factory because the SS came to arrest her. It was only after this that her role became somewhat exceptional, as she joined the armed struggle as the first female *gappista*[62] of Sesto San Giovanni. Small and extremely young-looking, she could pass unnoticed without arousing too much suspicion. Her main work was to transport bombs (as much as ninety kilos of explosives on one occasion), radios, and other materials, but she also took part in an action against the local German headquarters. In September

[61] On Vera Ciceri's Resistance activities see the biography of her husband Gaetano Invernizzi (Alasia, *Gaetano Invernizzi*).

[62] A member of a *gruppo d'azione patriottica* (GAP), the Communist Party's urban partisan action groups. These were strictly military clandestine organizations with many of their members living in hiding. Their role was to render life impossible for the German occupiers. They also had a protective role in defending mass actions such as strikes.

1944 she was arrested by the SS and taken to their headquarters in Monza.[63] After her release she continued to work clandestinely but with a new GAP, her old contacts being too dangerous.

Two other women who participated actively, if slightly less dramatically, were Anna Baroni and Teresina Ghioni. Anna Baroni from the N plant was one of the founders of the GDD in the firm. Her activities were somewhat bolder than those of other group members and on two occasions she addressed meetings flanked by an armed male partisan at each shoulder, first at the 'De Micheli' factory in Niguarda and then in the Fascist canteen in Piazza Maciachini.[64]

Teresina Ghioni was involved in antifascist activities from 1942 when she was only 15. As well as working with other general campaigns she specifically fought the introduction of night work for women, and campaigned against the transfer of machinery to Germany and against the deportation of women. She also helped recruit in the factory for the Garibaldi Brigade. In the strike of March 1944 she played a key role by turning off the electric current as the signal to down tools. Her own account of this is particularly interesting, as it typifies the unheroic way in which many female partisans described their contribution to the struggle:

> I was young, really young, but I managed to bring an entire factory to a halt, because I turned off the electricity at the mains. And they saw me! . . . They saw me, they were watching me . . . Just imagine! It was a really huge box where the mains switch was and there were lots of red buttons, I didn't know which one . . . I was lucky, because I didn't know which button, I went there and 'zum'! . . . They sent for me, dottore Ferrini [the head of the Personnel Office] sent for me and straight away said 'You!' and I said 'Not me'. I played the innocent, but I was trembling like a leaf, I must admit that I was terrified, because I

[63] Isa was arrested because she had arranged a meeting with Giovanni Pesce, the leader of her GAP, and they had been betrayed by fellow *gappista* 'Arconati'. She was, however, soon released for lack of concrete evidence as to her identity as a Communist. On the Resistance activities of Isa De Ponti see Alasia, *La vita di prima*, 189–201; Alasia, *Gaetano Invernizzi*, 119–21; G. Pesce, *Senza tregua: la guerra dei GAP* (Milan, 1967).

[64] She also openly refused to give funds to the Fascist party: G. Bianchi (ed.), *Dalla Resistenza: uomini, eventi, idee della lotta di Liberazione in provincia di Milano* (Milan, 1975), 169–70. Unfortunately I have no information on what happened to her as a result of her defiance.

was no hero: I did it, but afterwards I began to really think about it . . . You know that they used to take you to Germany and shut you away, tie you to the wall or put you in prison. I knew about those things. They let me go saying 'We'll hand you over to the German Headquarters', and I just replied in dialect; 'You can do what you want, I haven't done a thing!' I just went on saying 'Me? I haven't done a thing. What are you talking about? I don't know.' I just pretended I was thick. I was just a girl. I was only 17, but in those days 17-year-olds weren't what they are now.[65]

Teresina Ghioni, however, was in many ways an exception. Such dramatic activities as this, albeit unheroically described, were far from typical. Generally women's contributions were different and far less visible.

Much of the Resistance activities of women was organized by the GDD.[66] This group was the first to be set up in a Milanese factory. It was formed of women who had made contact during the 'forty-five days' but then spread via informal contacts in the workplace. Like the other GDD, the Magneti group contained women of varying political persuasions. Isa De Ponti described how the group came into being:

Vera (Ciceri) asked me to organize the women's group at Magneti. So I started by talking to girls that I knew well, who I could trust. With Anna Baroni—who they used to call Anna Cicietti because she was so thin—with her and with another woman who lived in Carugate we set up the first group. We called it *Iskra*, which means spark.[67] We collected

[65] By this pretence of innocence she went free. (Testimony of T. Ghioni interviewed by Luisa Finocchi.)

[66] These groups were important in coordinating women's role in the Resistance. The Magneti group was one of the pioneers, but soon after their numbers proliferated. From a mere 100 activists in Milan in April 1944 their numbers had swelled to about 3,400 by March 1945, with 184 groups, and a further 10,000 women linked to them. The national organization had its own press and included women from all parties and walks of life, ranging from workers and housewives to aristocrats, although they did tend to be mostly urban-based and dominated by Communists. As Angela Amoroso has pointed out, these groups must be seen as a landmark in women's history, since, despite their primary role as a mobilizer of women in supportive and nurturing roles, they also began to campaign around more feminist issues and to address the question of women's subordination in society. The issues they raised included the right to equal pay and training. (A. Amoroso, 'I gruppi di difesa della donna a Milano', paper presented at the conference 'Donna lombarda 1860–1945' (Milan, 1989).)

[67] *Iskra* is the Russian word for 'spark'.

things in the factory for the partisans. Clothes, warm things like coats: some of the women made socks, hats, and scarves by unravelling old clothes because it was impossible to find new wool in those days. We used to do this between midday and one o'clock, in our lunchbreaks as soon as we had finished eating; some of the women took things home to work on them in the evenings. One of them used to make wonderful woollen gloves—you know, it's difficult to knit gloves . . . When we got our pay packets we gave up a certain part of our money. We gave the things we had made or collected and the money to Vera.[68]

Women's role in the Resistance at Magneti, as organized by this group, therefore, was different from men's. Even in the Insurrection the Magneti women took a very specific role, as the following two contemporary accounts reveal:

Absolutely everyone, no matter what their job was in the firm—workers, engineers, clerks, or managers—they all took up arms at the Insurrection and acted as guards throughout the night: both men and women, *the latter making the food*.[69] (my italics)

Whilst the men are all engaged in military activity, the women from the *Gruppi di difesa*, joined by many Christian Democrat women, give their essential contribution to the struggle by looking after the provisions and assisting the fighters.[70]

Most women did not go beyond the rigidly defined norms of good female behaviour but participated primarily in a nurturing role. This was partly due to the emphasis on respectability and domesticity which had characterized their upbringing, and in view of the account of women's lives given in previous chapters it would have been very surprising if their Resistance role had been different. The period of the war, although in some ways emancipating for a section of the female population, was not enough to break down women's marginalization. As Isa said, 'It was the type of education they'd had, to be docile, always having to say yes, basically never rebelling because they weren't supposed to, and that's how they grew up.'

In addition, the women had to contend with the prejudices of

[68] 'La Narva' (testimony of Isa De Ponti), in Alasia, *La vita di prima*, 194.

[69] Valerio, 'Vita ed attività clandestina', 8.

[70] From E. Sereni, *Cln, il comitato di liberazione nazionale della Lombardia al lavoro nella cospirazione, nell'insurrezione, nella ricostruzione* (Milan, 1945). Reproduced in Vignati, *I ribelli al governo della città*, 69.

many male Resistance organizers. Resistance literature generally attempted to mobilize women as auxiliaries and only rarely addressed issues of female emancipation. Frequently male Resistance activists themselves failed to take women seriously and preferred to keep them in subordinate positions, as numerous clandestine documents demonstrate. The following extract which refers to Communist activity in the Sesto area is typical:

> The party organization has contributed little to the work amongst women, even now, after the excellent organizational successes, after the women have proved their ability to fight and sacrifice, often leading mass demonstrations, our male comrades sometimes prefer to ignore them.[71]

Given so many obstacles to their political participation, why then did some women get actively involved at all and, of these, why did some, like Isa De Ponti, carry bombs whilst others did nothing braver than knit socks?

There were three main political influences on these women's lives—factory, Church, and family. The political role of the firm as an agent of Fascist propaganda through the Dopolavoro and welfare institutions has already been examined. As I argued in the previous chapter, for some of the women that I interviewed the Church had, and still has to this day, an important impact on their lives. Religiosity varied. Some were not very keen churchgoers by the time they were workers in the firm, although religion had undoubtedly played a fairly significant role in their childhood lives. Religiosity seems to have declined for a variety of reasons. For Maria V., for example, it was the timetable of the factory which prevented her daily churchgoing. Others like Luciana stopped when they moved from the country to the city. Isa said that when she worked in a rural spinning mill 90 per cent of the women went to church every morning but at Magneti this was less common. But many of the Magneti

[71] 'Rapporto del Comitato femminile: rapporto sulla situazione organizzativa della federazione milanese, Zona A [Sesto San Giovanni], Milano 20 aprile 1944', cited in Scalpelli, *Scioperi e guerriglia,* 166. This problem dogged the activities of the GDD. One example is the fact that in June 1944 the GDD organization was officially recognized by the CLNAI. Despite this, 5 months later, representatives of the Lombardy GDD were still excluded from many key meetings for reasons of 'security'. (A. Amoroso, 'I gruppi di difesa'.)

women—whether urban or rural—were still very much practising Catholics. In rural areas the Church was important for most people, men or women. As Pierino said: 'Up there, in those villages . . . it's the priest who's in charge.' Where the Church moulded women's political ideas, it was a conservative influence, since the Vatican discouraged women's participation in public life.

In the formation of political ideas the Church was often subordinate to the family. A typical example was Carla, a resistance messenger, who described herself as 'born socialist'. Describing oneself as 'born socialist' or 'born Communist' seems to be common in oral testimonies relating to this period for both men and women.[72] It is evidence of the strength of commitment to certain political beliefs that they could be passed down from generation to generation even under Fascist repression. The family was crucial in this. Women who did get politically involved almost invariably formed their ideas at home, learning from other members of their families. Isa De Ponti herself was the daughter of a Communist worker from the Falck steelworks. She described the formation of her political beliefs: 'You could say that I was an antifascist and a Communist right from childhood.'[73] Teresina Ghioni said that she met antifascists on the shop-floor, but 'I knew about antifascism because I came from a family like that, so I knew that my father was an antifascist, I knew that my family didn't agree with certain things.'[74]

Given their narrow experience of life outside the home, in some ways it was inevitable that women would get involved via their families. Some male activists like Costantino continued the political tradition of their own families (Costantino's mother, father, aunts, and uncles had all been socialist maximalists), but it is also true that there was more chance that men from unpoliticized families might get involved. Giuseppe, for example, did not come from a political background, and Ercole recounted how he learned about antifascism through the workplace and friends.

Women, even more than men, were 'born' into politics,

72 See Passerini, *Fascism and Popular Memory*.
73 Interview with Isa De Ponti. Quoted in Alasia, *Gaetano Invernizzi*, 120.
74 Testimony of Teresina Ghioni.

because their lives were so strongly defined by their families. The public sphere generally had only a very limited presence in their lives. Factory life may have broadened their horizons to a certain extent, but in the end it was the family that was usually the primary influence on the formation of political ideas. This, of course, was limiting for women, since it inevitably meant that less women than men got involved, particularly as it was precisely the family which was so deeply involved in safeguarding women's morality and limiting their ability to participate in the outside world.

The fact that some women, like Isa De Ponti, were so active is especially impressive in view of the numerous obstacles to women's participation in the public sphere at all. Dominated as their lives were by family life, often burdened by housework and restricted by strong notions of respectability, most women had neither the time nor the freedom to do anything more than make clothes for the partisans in their lunchbreaks.

A DIFFERENT KIND OF RESISTANCE

This different role of women in the Resistance may go a long way towards explaining why far fewer workers were deported or killed in this firm than in others like Breda. The evidence to prove that the work-force was somehow more conservative, or that the Nazi and German repression was deliberately less, is inconclusive. Although only the most tentative conclusions can be offered on the political attitudes of the Magneti workers, according to that information I have been able to find, Magneti does not seem to have been a local Mecca for Fascists, despite management claims to the contrary. It is quite possible that, at Magneti, the numbers of workers involved in the Resistance equalled those in other firms, except that here the sort of antifascist activity that was going on was different. About half of the work-force was female, and although many women got involved, few took up arms or became organizers. Those who did got a certain amount of protection from arrest from the lesser suspicion they aroused.

Nevertheless, the Resistance period and the fall of Fascism must have marked a watershed in the lives of many of these women. The description of their daily lives in previous chapters reveals a very private world. In spite of the fact that only a small percentage of the Magneti women really got involved in activities which could have led them to question women's traditional role, even most of those who did not follow in Isa De Ponti's footsteps took part in the strikes. For them this period was one where the protective shell of family and factory could no longer totally separate them from the public sphere.

Conclusion

FOUNDED as a result of the demand for arms in the First World War, by the fall of Mussolini's Fascist regime Magneti Marelli was a large and thriving concern. One of the reasons for such rapid growth was undoubtedly the fact that it was one of the first firms in Italy to introduce 'Americanized' production methods, enabling it to compete with advanced foreign firms. In contrast to many other Italian firms where 'Americanization' came to mean little more than 'speed-up', Magneti was one of the few firms to introduce Taylor's and Ford's ideas on a large scale, even extending its desire to be 'scientific' to aspects such as recruitment and industrial health following the ideas of the revisionist Taylorites. In this respect, Magneti was one of the most modern firms in inter-war Italy. Yet the history of this firm was also marked by the archaic, reactionary, and contradictory nature of the Fascist period, as some of the Dopolavoro activities and the particular conditions in Luigia Pedroni's Winding Shop demonstrate.

In the absence of effective unions, and faced by authoritarian and hierarchical organizational structures within the factory, the work-force could do little to control or influence the way in which Taylorism was introduced, except by using secret 'tricks' to keep the bonus rate up. The pace of work was intense and many jobs were monotonous and repetitive. One effect of the organization of the work process was that there were frequently problems of jealousy amongst the women, since the condition of work made them insecure and competitive. Between the sexes, on the other hand, there was a marked absence of gender conflict, largely due to the fact that male and female roles on the shop-floor were very strictly defined.

The Taylorist-Fascist ideal that workers and employers could co-operate happily in a new Italy needed a more effective advocate than Fascist cudgels and rigid shop-floor discipline if workers were to accept that the management really 'knew best'. The management, therefore, made use of more subtle methods, ranging from the publication of a factory newspaper to the provision of a wide range of welfare facilities. Most of the latter involved a certain amount of Fascist rhetoric; some were supposedly a part of the rationalization of production, whilst others were simply classic paternalism.

Magneti Marelli was in many respects atypical of inter-war Italian industry. It was unusually modern in its technology and management methods, and was one of the first Italian firms to have a fairly stable labour force. The history of such an unusual and innovative firm merits study, if only to further historical knowledge of the spread of scientific management and the impact of this upon the working class as a whole. However, this firm was also a large employer of women and, as such, provides a thought-provoking case study into the female experience of the Fascist period.

The female half of the work-force experienced employment in this firm in a very different way from men. Indeed, although in general there was a lack of actual spatial segregation (except for the winders), it is still possible to talk of veritable male and female 'spheres' within the firm. The female 'sphere' was characterized by lower pay, lower status, and lighter, more monotonous work roles. Despite this, once they had obtained work here, women tried to cling on to it as, in contrast to many other types of employment open to them locally, working in this firm was undoubtedly attractive. This factor at least partly explains why labour mobility differed between men and women, and why average turnover was relatively low overall in this firm. The scarcity of good jobs for women and the impact of welfare provisions such as the nursery made them a stable and potentially loyal section of the work-force.

This case study shows one of the many faces of the female experience under Fascism. Clearly these women failed—together with large numbers of the rest of the female Italian population—to match up to Mussolini's monolithic ideal of true Fascist womanhood. Ideological myths about 'true femininity' and the generally repressive atmosphere of the period undoubtedly did have some effect on their lives, but it is evident that in many respects these women's lives were affected as much, if not more, by factors which cannot be seen as specifically Fascist, such as the pattern and manner of industrialization and urbanization.

Chiara Saraceno has argued that because of these two factors, this period—in spite of Fascist misogynist rhetoric—may have been one of relatively increasing 'emancipation' and greater equality between the sexes for urban working-class couples.[1] The example of the Magneti women does at least in part bear out Saraceno's positive assessment. Urbanization and factory work do appear to have led to some degree of female emancipation. In the teeth of Fascist ideology condemning female waged employment outside the home, they had a strong commitment to the value and dignity of work and forged for themselves, despite the monotonous nature of the tasks they did, a clear self-image as workers. It is likely that many were practising at least some sort of control of their fertility, and access to a workplace nursery meant that it was possible to combine employment with maternity. Within the home some degree of equality may have been emerging. For those who married, there is some evidence that in certain cases husbands did help with at least the lighter housework and occasionally shared many of the household duties.

On the other hand such 'emancipation' had many limitations. The potential challenge posed to men's dominance by female waged employment outside the home was to a great extent offset by the patriarchal organization of the workplace. Women were only tolerated in the hitherto masculine preserve of engineering if they did not threaten the male sphere of better-paid skilled work. They were stuck in what were basically dead-end jobs with only extremely limited chances of promotion. In the firm's initiatives which were aimed particularly at women the emphasis was firmly on domesticity and gendered notions of respectable behaviour. Furthermore, these women's own lack of confidence in the workplace led them in certain cases to accept and even prefer restrictions on their freedom. For example, the oppressive atmosphere of the Winding Shop could seem a safe haven from the mixed environment of the other shops where sexual harassment might threaten. Wages, even though better than those which could be commanded by

[1] C. Saraceno, 'Percorsi di vita femminile nella classe operaia: tra famiglia e lavoro durante il fascismo', *Memoria*, 2 (1981).

women in certain other jobs, were still far too low for them even to consider as a positive choice an independent life without the protection of a male breadwinner. For female heads of families, life was an uphill, difficult struggle. For some of these women, even those who lived in urban areas, the Church was still a powerful influence on their lives and, despite some degree of male help, their home lives remained burdened by housework, rendering leisure time virtually non-existent. In the sphere of politics, whether Fascist or antifascist, most women had only a limited and gender-specific role. Their participation in the Fascist organization to 'nationalize the masses' was marginal. In the antifascist Resistance most of the handful of women who took a particularly active role did so because of political socialization in left-wing families, rather than as a result of the emancipating impact of waged work outside the home. The role of the other women who got involved remained an auxiliary one and rarely presented a challenge to prevailing social norms of respectable female behaviour.

But to evaluate women's experience of this period in terms of how 'emancipating' or 'repressive' it was for them is, of course, a simplification. Although these women did not encroach much on the male sphere and gender divisions remained safeguarded (if somewhat renegotiated), within their subordinate sphere women were not simply passive victims of their situation. They may have been largely excluded from the Dopolavoro, but women-only non-work 'leisure' activities did exist in the shape of needlework, a creative craft which was a source of both pleasure and pride to them. Likewise, women's contribution to the Resistance is better understood on its own terms. Rather than attempting to assess the extent of their contribution to male-dominated activities, it is more useful to focus on women's specific and separate role. This shift of perspective highlights the fact that female activities in the resistance period were, in fact, substantial.

Although the case study focused on in this book is of a somewhat exceptional factory, many of the experiences of the women who worked there may not have been particularly unusual. Such aspects as the strict gendering of work roles, the absence of ambition, the work ethic, and the relationship between home and work and its impact on leisure activities may all have been fairly widespread in the industrial north of Italy,[2] although only further research can determine the extent to which this was so.

Whether typical or not, the example of the Magneti Marelli work-force suggests that historians should treat more carefully certain criteria which have been used to assess the history of the working class in this period. For example, I hope I have successfully demonstrated that the emergence of 'modern, stable, industrial labour forces' was not simply determined by factors affecting patterns of male employment, as some recent studies seem to imply, and that this phenomenon can only be properly understood in the context of gender divisions. The unusual stability of the Magneti labour force can be attributed to a series of factors, but the most important of these was the particular situation of women in the labour market. Similarly, the example of the Magneti Dopolavoro shows the care historians must take in assessing the impact of Mussolini's policies to 'nationalize the masses'. A study of women's leisure activities highlights the fact that the question of popular leisure and worker's free time under Fascism is by no means exhausted by a study of the Dopolavoro, since, for half of the working class, it was a virtually irrelevant institution.

With regard to this, as with many other aspects of the Fascist experience in Italy, it is evident that only by taking into account the gendered nature of inter-war Italian society can historians hope to reach a fuller understanding of this period. Working-class women did share some of the same problems as men, and arguably may have been as much affected by policies directed to suppress working-class autonomy as by the misogyny of Fascist ideology and policies. However, working-class women's experience was in many respects quite distinct. An account of their lives provides both a challenge to and an enrichment of the traditionally male-dominated historiography of the Italian working class.

[2] The society of the south of Italy was so different in this period both culturally and economically that there is no reason to suggest particular similarities of experience.

APPENDIX 1: *Interviewee profiles*

Name	Date of birth	Place of birth	Residence in 1930s	Date of marriage	Started at FIMM	Shop/job
1. Bruna	1921	Parma	Milan	1947	1938	Winding
2. Carla	1910	Lecco	Sesto	1931	1925	Winding
3. Delfina	1919	Sesto	Sesto	1939	1935–40, 1942–7	Radio
4. Germana	1909	Ferrara	Sesto	1939	1927	Radio
5. Gina	1901	Vaprio	Sesto	1919, 1938	1935?	Magnetos
6. Giuseppina	1921	Bergamo	Sesto/Monza	197?	1937	Winding/Radio/ Sparkplugs
7. Isa	1920	Sesto	Sesto	—	1940	Pay clerk
8. Luciana	1921	Rovigo	Sesto	1942	1938	Sparkplugs
9. Lucina	1924	Sesto	Sesto	1948	1939	Chemical
10. Luigia	1924	Sesto	Sesto	—	1938	Winding/chemical
11. Margherita	1907	Como	Sesto	—	1922	Magnetos/nursery nurse
12. Maria B.	1911	Parma	Milan	—	1930?	Pay clerk
13. Maria G.	1917	Rovigo	Monza	1942	1933	Winding
14. Maria V.	1919	Como	Sovico	1949	1935	Winding
15. Olga	1915	Sesto	Sesto	1938	1932	Cutting/Galvanizing
16. Orsola	1909	Cremona	Sesto	1933	1935	Radio
17. Pierina	1923	Sesto	Sesto	1950	1938	Winding

18. Costantino	1915	Turin	?	1941	1937	Op. machines/ Toolroom
19. Dino	1921	Monza	Monza	1961	1934	Toolroom/Technical Office
20. Enrico	1921	Sesto	Sesto	1949	1935	Toolroom
21. Ercole	1927	Milan	Milan	1953	1941	Toolroom
22. Ermano	1923	Friuli	Milan	1955	1938	Draftsman
23. Gaetano	1911	Bari	Sesto	1931	1927	Testing room
24. Giglio	1923	Ferrara	Sesto	1949	1938	Toolroom
25. Giovanni	1905	Monza	Monza	1935	1941	Toolroom
26. Giuseppe M.	1927	Novara	Milan	1958	1941	Toolroom/Screw-cutting
27. Giuseppe R.	1915	Milan	Milan	1945	1937	Radio
28. Luciano	1924	Milan	Milan	1955	1938	Technical office
29. Luigi B.	1914	Rovagnate	Rovagnate	—	1935	Winding
30. Luigi C.	1915	Sesto	?	1946	1933	Technical office/clerk
31. Paride	1912	Milan	Milan	1938	1935	High Altitude Magnetos
32. Pietro	1915	Bergamo	?	1944	1934	Labourer, B Plant
33. Santino	1905	Sesto	Sesto	?	1935	V section
34. Teodoro	1909	Monza	Monza	1946	1939	Brakes

APPENDIX 2:
Additional tables on the composition of the work-force

SOME METHODOLOGICAL NOTES ON THE TABLES PRESENTED HERE AND IN CHAPTER 5

The detailed personnel records of the Magneti Marelli archive constitute a very rich source of information for understanding the composition of the labour force. There are, however, certain problems inherent in the use of this type of material. Unlike the payroll registers used in most other analogous studies, these personnel files are not kept in bound books but in large bundles simply tied together with string. Thus it is far more likely that some may have been lost or destroyed. Without the payroll registers it is very difficult to cross-check this. A second problem lies in the fact that, as Table P shows, the recording of data was very uneven, so that some of the files contain only minimal information such as name, date of birth, etc. The files with the least information are often those of workers who were recruited in the 1920s. This probably reflects the more efficient organization of the Personnel Department in the later period. Nevertheless the unusually wide range of data contained in this archive makes it valuable in spite of these problems. Where I consider that these problems are particularly acute I have made this clear in the text.

Not all this huge amount of data could be analysed quantitatively and I decided to exclude certain types of information such as that on pay, which was complex, uneven, and not easily classified. However, I recorded the greatest part and, after classification and coding, the computer analysis included both the original variables and others which were computed from these to give, for example, the worker's age at recruitment and the length of time spent employed at Magneti. In the process of analysis, however, some of the information proved to be unusable, such as the data about the presence of other members of the worker's family in the factory.

The sample was composed of the data from the file of every fourth woman and every twelfth man who started work at Magneti up to 10 June 1940. The differing proportions of men and women taken were due to the fact that my original plan was to use the female sample for cross-linking with the local municipal birth, death, and marriage records in order to examine the interaction between factory work and family for the women. I wished, therefore, to have a sizeable female sample. Unfortunately, my original project proved far too ambitious and would have required an enormous amount of additional work beyond the scope of a single researcher working alone (not least due to the fact that the *fogli di famiglia* have been lost in the Sesto municipal archives and matching up the data would have been a huge task), and so my plans had to be shelved. To make comparisons easier, figures for the male sample are weighted by a factor of three in the tables. Overall, the sample included 1,023 files of which 635 belonged to female workers and 387 to males (and one of unknown sex!).

The sampling method of taking every *n*th worker was chosen, rather than using only certain bundles of folders which would have been quicker, because the archive is arranged alphabetically. It seemed important to avoid the predominance of certain surnames which could have severely distorted the information on the geographical origin of the workers.

Figures in all tables refer to the sample (multiplied by 3 in the case of men), not the entire work-force.

Table A Sex of recruits by year of starting work

Year started work	Women		Men		Total
	No.	%	No.	%	
1911–25	32	40.0	48	60.0	80
1926–30	50	31.0	111	68.9	161
1931	4	10.0	36	90.0	40
1932	39	52.0	36	48.0	75
1933	67	42.7	90	57.3	157
1934	27	22.5	93	77.5	120
1935	112	40.4	165	59.6	277
1936	84	42.4	114	57.6	198
1937	70	29.0	171	71.0	241
1938	20	27.0	54	73.0	74
1939	94	32.5	195	67.5	289
1940	33	44.0	42	56.0	75
Total	632		1,155		1,787

Table B Initial grades of recruits to Plants A and B

	Plant A		Plant B		Total
	No.	%	No.	%	
Female B grade	62	15.4	—	—	62
Girl	27	6.7	1	0.7	28
Female apprentice	82	20.4	2	1.4	84
Operaio specializzato	42	10.4	—	—	42
Operaio qualificato	39	9.7	3	2.1	42
Male apprentice	18	4.5	18	12.8	36
Manovale specializzato	60	14.9	63	44.7	123
Manovale comune	27	6.7	48	34.0	75
Boy	45	11.2	6	4.3	51
Total	402		141		543

Table C Initial grades

WOMEN Grade	1911–30		1931–5		1936–40		Total	
	No.	%	No.	%	No.	%	No.	%
A grade	—	—	—	—	1	0.6	1	0.3
B grade	5	17.9	3	2.5	124	77.0	132	42.5
Girl	6	21.4	17	13.9	19	11.8	42	13.5
Apprentice	17	60.7	102	83.6	17	10.6	136	43.7
Total	28		122		161		311	
MEN								
Operaio specializzato	9	23.0	15	6.2	57	17.8	81	13.5
Operaio qualificato	12	30.8	33	13.8	30	9.3	75	12.5
Apprentice	—	—	42	17.5	21	6.5	63	10.5
Manovale specializzato	—	—	54	22.5	144	44.9	198	33.0
Manovale comune	12	30.8	57	23.8	30	9.3	99	16.5
Boy	6	15.4	39	16.2	39	12.2	84	14.0
Total	39		240		321		600	

The first two grades for men (*operaio specializzato* and *qualificato*) are the most highly skilled grades, whereas the unskilled and semi-skilled were generally classified as *manovali*.

Table D Educational level of recruits according to year started work

WOMEN Year started work	Primary		Secondary		Total
	No.	%	No.	%	
1911–30	55	93.2	4	6.8	59
1931–5	106	86.2	17	13.8	123
1936–40	80	93.2	6	7.0	86
Total	241		27		268
MEN					
1911–30	72	63.2	42	36.8	114
1931–5	144	57.1	108	42.9	252
1936–40	99	40.7	144	59.3	243
Total	315		294		609

Table E Level of education of male workers by initial grade

Grade	Primary		Secondary		Total
	No.	%	No.	%	
Operaio specializzato	24	42.1	33	57.9	57
Operaio qualificato	39	72.2	15	27.8	54
Apprentice	18	50.0	18	50.0	36
Manovale	96	72.7	36	27.3	132
Boy	27	37.5	45	62.5	72
Total	204		147		351

Table F Parental status of married recruits

	Women		Men	
	No.	%	No.	%
Childless	52	53.1	102	27.0
With children	46	46.9	276	73.0
Total	98		378	

Table G Reasons for leaving work

Reason for leaving	Women		Men		Total
	No.	%	No.	%	
Maternity	35	29.4	—	—	35
Family reasons	8	6.7	3	1.0	11
Ill health	14	11.8	27	9.2	41
Absenteeism	4	3.4	3	1.0	7
Disciplinary reasons	3	2.5	6	2.1	9
Old age or low output	9	7.6	36	12.2	45
Reduction of work-force	6	5.0	36	12.2	42
Left of own will	7	5.9	33	11.2	40
Events of war*	3	2.5	42	14.3	45
Internal transfer†	13	10.9	48	16.3	61
Death	2	1.7	18	6.2	20
Sacked to be rehired immediately‡	15	12.6	42	14.3	57
Total	119		294		413

* Joined armed forces or purged as Fascist.
† Promoted to clerical grade, or transferred to another plant.
‡ Sometimes this was to reclassify the worker in order to cut pay.

Table H Mean age of women workers when they left the factory according to year of recruitment

Year started work	Mean age (in years)	No. of cases	Standard deviation
1927	28.8	7	9.2
1928	28.6	12	11.5
1929	33.1	4	6.7
1930	30.7	17	7.1
1931	31.2	3	11.3
1932	26.4	23	4.8
1933	24.1	37	5.0
1934	27.1	24	4.3
1935	28.5	97	8.8
1936	27.1	80	6.3
1937	27.5	70	8.7
1938	24.1	20	8.3
1939	25.9	90	6.3
1940	27.2	30	8.3

Table I Place of birth according to year of starting work

Place of birth	Up to 1930		1931–5		1936–40		Total	
	No.	%	No.	%	No.	%	No.	%
Sesto San Giovanni	32	15.5	53	8.3	51	6.6	136	8.4
City of Milan	33	15.9	56	8.8	68	8.9	157	9.7
Prov. of Milan	45	21.7	211	33.1	219	28.5	475	29.4
Rest of Lombardy	17	8.2	38	6.0	86	11.2	141	8.7
Bergamo, Brescia, Como	11	5.4	84	13.2	115	15.0	210	13.0
Emilia and Venetia	47	22.7	99	15.5	121	15.8	267	16.6
Piedmont and Liguria	3	1.4	29	4.5	15	2.0	47	2.9
Centre and South	13	6.3	46	7.2	88	11.5	147	9.1
Abroad	6	2.9	22	3.4	5	0.7	33	2.0
Total	207		638		768		1,613	

Table J Place of residence according to year of starting work

WOMEN Place of residence	Up to 1930		1931–5		1936–40	
	No.	%	No.	%	No.	%
Sesto San Giovanni	54	78.3	126	52.9	109	47.6
Adjacent towns	13	18.8	56	23.5	46	20.1
Monza	—	—	26	10.9	33	14.4
5–10 km from factory	—	—	3	1.3	6	2.6
10–15 km from factory	—	—	4	1.7	5	2.2
Over 15 km from factory	1	1.4	15	6.3	17	7.4
City of Milan	1	1.4	8	3.4	13	5.7
Total	69		238		229	
MEN						
Sesto San Giovanni	45	32.6	99	24.1	111	20.1
Adjacent towns	27	19.6	69	16.8	87	15.8
Monza	21	15.2	57	13.9	63	11.4
5–10 km from factory	—	—	9	2.2	21	3.8
10–15 km from factory	—	—	12	2.9	9	1.6
Over 15 km from factory	9	6.5	63	15.3	99	17.9
City of Milan	36	26.1	102	24.8	162	29.4
Total	138		411		552	

Table K Length of employment by place of residence

WOMEN Place of residence	Up to 6 months		6–12 months		1–2 years		2–5 years		Over 5 years		Total
	No.	%	No.	%	No.	%	No.	%	No.	%	
Sesto San Giovanni	18	7.0	19	7.5	20	8.0	63	25.0	133	52.5	253
Adjacent towns*	7	7.4	2	2.1	13	26.3	25	26.3	48	50.5	95
Monza	3	6.1	4	8.2	7	14.3	21	42.8	14	28.6	49
5–10 km away	—	—	—	—	—	—	5	62.5	3	37.5	8
10–15 km away	1	12.5	1	12.5	—	—	5	62.5	1	12.5	8
Over 15 km away	—	—	6	20.7	2	6.9	10	34.5	11	37.9	29
City of Milan	3	16.7	1	5.5	3	16.7	4	22.2	7	38.9	18
Total	32		33		45		133		217		
MEN											
Sesto San Giovanni	39	17.8	30	13.7	33	15.1	63	28.8	54	24.6	219
Adjacent towns*	24	15.4	9	5.8	30	19.2	27	17.3	66	42.3	156
Monza	9	7.9	18	15.7	12	10.5	21	18.5	54	47.4	114
5–10 km away	6	25.0	3	12.5	—	—	12	50.0	3	12.5	24
10–15 km away	3	20.0	—	—	—	—	6	40.0	6	40.0	15
Over 15 km away	24	16.6	12	8.3	27	18.8	45	31.3	36	25.0	144
City of Milan	51	21.2	30	12.5	39	16.3	69	28.8	51	21.2	240
Total	156		102		141		243		270		

* Cologno Monzese, Cinisello, Bresso, Vimodrome, Brugherio, Segrate, and the zone of Milan next to the factory.

Table L Monthly pattern of recruitment

Month started work	Women		Men		Total	
	No.	%	No.	%	No.	%
January	31	5.8	39	4.0	70	4.6
February	13	2.4	60	6.1	73	4.8
March	11	2.0	66	6.7	77	5.1
April	9	1.7	66	6.7	75	5.0
May	6	1.1	60	6.1	66	4.4
June	22	4.1	78	8.0	100	6.6
July	46	8.6	99	10.1	145	9.6
August	60	11.2	78	8.0	138	9.1
September	72	13.4	93	9.5	165	10.9
October	113	21.0	129	13.2	242	16.0
November	131	24.4	141	14.4	272	18.0
December	23	4.3	69	7.1	92	6.1
Total	537		978		1,515	

Table M Monthly pattern of leaving work

Month left work	Women		Men		Total	
	No.	%	No.	%	No.	%
January	17	3.6	51	6.4	68	5.4
February	35	7.5	78	9.8	113	9.0
March	44	9.4	117	14.7	161	12.8
April	31	6.7	75	9.4	106	8.4
May	52	11.2	57	7.2	109	8.6
June	41	8.8	63	7.9	104	8.2
July	39	8.4	69	8.7	108	8.6
August	34	7.3	39	4.9	73	5.8
September	55	11.8	51	6.4	106	8.4
October	50	10.7	63	7.9	113	9.0
November	34	7.3	75	9.4	109	8.6
December	34	7.3	57	7.2	91	7.2
Total	466		795		1,261	

Table N Length of employment by initial grade

Grade	Up to 6 months		6–12 months		1–2 years		2–5 years		Over 5 years		Total
	No.	%	No.	%	No.	%	No.	%	No.	%	
Woman B grade	2	1.6	2	1.6	19	15.3	65	52.4	36	29.1	124
Girl	2	5.9	2	5.9	4	11.8	10	29.4	16	47.0	34
Apprentice	2	1.8	5	4.5	2	1.8	10	8.9	93	83.0	112
Total women	6		9		25		85		145		270
Operaio specializzato	3	5.0	3	5.0	9	15.0	30	50.0	15	25.0	60
Operaio qualificato	—	—	—	—	3	5.0	18	30.0	39	65.0	60
Apprentice	12	26.7	6	13.3	9	20.0	18	40.0	—	—	45
Manovale specializzato	15	8.0	24	12.7	30	15.9	36	19.0	84	44.4	189
Manovale comune	15	20.8	12	16.7	9	12.5	21	29.2	15	20.8	72
Boy	6	8.3	6	8.3	12	16.7	36	50.0	12	16.7	72
Total men	51		51		72		159		165		498

Table O Length of employment according to year of starting work

WOMEN Length of employment	1911–30		1931–5		1936–40		Total
	No.	%	No.	%	No.	%	
Up to 1 month	2	2.6	2	1.1	6	2.1	10
1–6 months	6	7.7	7	3.8	29	10.0	42
6–12 months	2	2.6	15	8.0	31	10.6	48
1–2 years	2	2.6	15	8.0	48	16.5	65
2–5 years	8	10.2	44	23.7	104	35.7	156
5–10 years	11	14.1	82	44.1	57	19.6	150
10–20 years	30	38.4	17	9.1	15	5.2	62
20–39 years	17	21.8	4	2.2	1	0.3	22
Total	78		186		291		555
MEN							
Up to 1 month	6	4.0	9	3.1	42	8.1	57
1–6 months	9	6.0	27	9.2	66	12.7	102
6–12 months	12	8.0	36	12.2	66	12.7	114
1–2 years	15	10.0	39	13.3	93	17.9	147
2–5 years	18	12.0	78	26.5	153	29.5	249
5–10 years	24	16.0	57	19.4	57	11.0	138
10–20 years	42	28.0	39	13.2	39	7.5	120
20–39 years	24	16.0	9	3.1	3	0.6	36
Total	150		294		519		963

Table P Data initially recorded

	No. of files on which information is recorded
Sex	1,022
Date of birth	1,016
Place of birth	887
Place of residence	909
Plant in which employed	569
Marital/parental status	748
Fascist Party membership	27
Date of first recruitment	1,018
Date of first leaving	883
Reason for leaving	197
Initial grade	517
Final grade	475
Initial job	608
Final job	320
Members of family in factory	233
Level of education	474
Previous work experience	569

Total number of cases = 1,023.

APPENDIX 3:
Population growth and the labour market in Sesto San Giovanni

Table 1 Population of Sesto San Giovanni

	Present	Resident
1901 census	7,021	6,952
1911 census	14,138	15,807
1921 census	18,864	18,274
1931 census	31,044	30,568
31 Aug. 1934	36,214	35,845

Year	Births	Deaths	Immigrants	Emigrants
1920	494	285	1,564	706
1921	506	271	262	583
1922	509	277	817	589
1923	483	275	780	535
1924	436	256	1,074	655
1925	443	289	1,502	645
1926	458	334	2,328	760
1927	531	323	2,481	832
1928	587	330	1,921	816
1929	630	339	3,152	1,009
1930	675	316	3,369	698
1931	747	390	3,603	2,069
1932	755	410	2,788	2,682
1933	704	375	2,609	1,628
Total	7,958	4,470	28,750	14,207

Source: Fascio di Combattimento di Sesto San Giovanni, *Sesto San Giovanni fascista industriale* (Milan, 1934).

Table 2 Pattern of male and female industrial employment in Sesto San Giovanni according to the 1921 census

	Men	Women	Total
1. Metallurgy and engineering	2,847	152	2,999
2. Mineral processing and building	1,088	165	1,253
3. Textiles	187	927	1,114
4. Woodworking	500	27	527
5. Transport	460	11	471
6. Industries using animal and vegetable products	222	28	250
7. Chemical industry	118	130	248
8. Production and distribution of energy	169	8	177
9. Printing	60	22	82
10. Other industries	57	68	125
Total	5,708	1,538	7,246

APPENDIX 4: *Pay tables*

All the following tables have been compiled from the personnel records of the factory, some of which contain data on the level of pay. The information given is complex. It does not refer to the actual pay a certain worker received on a specific day, but is a kind of check designed to see how much each he or she was earning on average. A sample time period was taken, such as a week or a month, or more usually a few months, and the number of hours worked noted. The base pay per hour was also noted. From this an average per day and per hour was calculated.

All figures are in lire.

Table 1 Pay of Ester (started work 1917)

Year	Base pay per hour	Average pay per hour	Pay per day	Grade/job
1924	1,35		14,56	
1925	1,35		16,00	
1926	1,40		18,80	
1927	1,45		19,10	
1928	1,45		19,10	
1929	1,50		19,10	
1930	1,50		24,00	Assembly worker
1931	1,38			
1932	1,38	2,00	16,00	
1933	1,38	2,04	16,70	
1934	1,28	1,94	15,50	
1935	1,28	1,99	15,92	
1936	1,90	2,04	16,32	
1937	1,90	2,86	22,88	Woman A
1938	2,09	2,98	23,84	
1939	2,30	3,04	24,32	
1940	2,65	3,55	28,40	
1941	2,65	3,91	31,28	
1942	2,65	4,16	33,88	

Table 2 Pay of Antonia (started work 1920)

Year	Base pay per hour	Average pay per hour	Pay per day	Grade/job
1923	1,10			
1924	1,10		7,40	
1925	1,10		11,30	
1926	1,26		18,40	
1927	1,35		18,40	
1928	1,35		18,30	
1929	1,40		14,00	
1930	1,40		14,00	
1931	1,20			
1932	1,28	2,00	16,00	
1933	1,28	2,11	17,20	
1934	1,19	1,88	15,04	
1935	1,19	2,34	18,72	
1936	1,19	2,59	20,70	
1937	1,77	3,03	24,24	
1938	1,95	3,13	25,04	Woman A
1939	2,15	3,34	26,72	
1940	2,47	4,39	35,12	
1941	2,47	2,94	23,52	
1942	2,47	2,53	20,24	
1943	2,47	2,51	20,08	

Table 3 Pay of Guidetta (started work 1928)

Year	Base pay per hour	Average pay per hour	Pay per day	Grade/job
1928	0,90		7,50	
1929	0,90		8,30	
1930	0,90		8,30	
1931	0,95			
1932	0,87	1,38	11,04	
1933	0,87	1,33	10,64	
1934	0,83	1,34	10,72	
1935	0,81	1,39	11,12	
1936	0,81		10,96	
1937	1,32	1,57	12,56	
1938	1,32	1,70	13,60	
1939	1,54	2,00	16,00	
1940	1,84	2,96	23,68	
1941	1,84	2,72	21,76	

Table 4 Pay of Rosa (started work 1928)

Year	Base pay per hour	Average pay per hour	Pay per day	Grade/job
1928	0,80		12,00	Apprentice
1929	0,80		12,00	
1930	0,90			
1931	0,82			
1932	0,82	1,16	9,28	Shearer
1933	0,82	1,19	9,52	
1934	0,76	1,22	9,76	
1935	0,76	1,42	11,36	
1936	0,76	1,42	11,36	
1937	1,13	1,44	11,52	Woman B
1938	1,24	2,68	13,44	
1939	1,36	2,05	16,40	
1940	1,56	2,55	20,40	
1941	1,56	2,59	20,70	

Table 5 Examples of pay of various women in their first year of work

Year	Base pay per hour	Average pay per hour	Pay per day	Grade/job
1922	0,90			Polisher
1923	1,20			Tester
1924	1,00		7,20	Welder
1925	1,10		15,30	
1926	1,20		15,50	
1927	0,90		13,20	Polisher
1928	0,90		13,00	Drill operator
1929	0,90			
1930	0,90			Assistant
1931	0,87			Winder
1932	0,82	1,16	9,28	Shearer
1933	0,82	1,60	12,80	Assistant tester
1934	0,76	1,06	8,48	
1935	0,76	1,63	13,04	
1936	0,76	1,55	12,40	Assistant
1937	1,24	1,49	11,9	Woman B
1938	1,24	2,04	16,64	Woman B
1939	1,36	1,84	14,72	Woman B
1940	1,56	1,99	15,92	

Table 6 Pay of Alberto (started work 1919)

Year	Base pay per hour	Average pay per hour	Pay per day	Grade/job
1925	2,50		41,00	Fitter
1926	2,50		43,00	Foreman
1927	2,70		34,40	
1928	2,80		41,00	
1929	2,80		32,65	
1930	2,85		34,00	
1931	2,62			
1932	2,62	4,00	32,00	Assembler-fitter
1933	2,62	3,66	29,32	
1934	2,62	3,63	29,04	
1935	2,44	3,80	30,40	
1936	2,44	3,83	30,64	
1937	3,63	4,34	39,02	
1938	3,99	5,57	44,56	Semi-skilled worker
1939	4,39	5,63	45,04	Qualified worker
1940	5,05	6,40	51,20	
1941	5,05	7,36	58,88	
1942	5,05	7,61	60,88	
1943	5,05	8,13	65,04	

Table 7 Pay of Carlo (started work in 1919)

Year	Base pay per hour	Average pay per hour	Pay per day	Grade/job
1924	1,95		33,17	Labourer
1925	1,95		32,24	
1926	2,05		30,00	
1927	2,15		26,00	
1928	2,20		34,00	Sander
1929	2,25		34,00	
1930	2,25			
1931	2,07			Labourer
1932	2,07	2,80		Toilet cleaner
1933	2,07		24,00	
1934	1,93	2,62	20,96	
1935	1,93	2,99		
1936	1,93	2,76	22,08	
1937	2,86	2,98	23,84	
1938	3,15	3,30	26,40	Ordinary labourer
1939	3,47	3,40	27,20	
1940	3,99	4,04	32,32	
1941	3,99	4,19	33,52	
1942	3,99	4,17	33,36	
1943	3,99	4,82	38,56	
1944	3,99			

APPENDIX 5:
Initial jobs done by Magneti workers in sample

Women	Men	Job
1	0	Aligner
3	0	Attendant
32	27	Apprentice
16	54	Assembly worker
0	9	Automatic machine operator
100	69	Auxiliary
0	9	Bricklayer/mason
0	6	Buffer/polisher
0	3	Cabinetmaker
3	0	Canteen worker
0	3	Caster
1	0	Cook
0	3	Draughtsman
7	6	Drill operator
0	9	Driver
0	36	Electrician
0	6	Enamellist
0	3	Filer
0	135	Fitter
0	12	Foreman
0	3	Forger
0	39	Foundry worker
0	3	Gauger-fitter
13	9	General worker
4	9	Grinding machine operator
0	3	Hammerman
7	30	Inspector/checker
0	15	Joiner
0	12	Labourer
0	3	Maintenance worker
0	21	Mechanic
0	3	Mechanic/engineer
0	3	Messenger
2	18	Milling machine operator
1	0	Nursery nurse
0	3	Operator
2	0	Overlooker
0	6	Plumber

Women	Men	Job
0	3	Polisher
0	60	Pug mill operator/mixer
0	6	Radio assembly worker
1	24	Radio engineer
0	3	Radio repair mechanic
0	3	Repair worker
10	6	Shearer
0	21	Smith
1	0	Tapping machine operator
1	0	Telephonist
1	0	Tester/inspector
0	27	Toolmaker
0	18	Toolmaker-fitter
1	3	Trimmer/snagger
3	120	Turner
1	3	Varnisher/painter
2	12	Warehouse worker
1	15	Welder
68	0	Winder
17	3	Worker

Source: APMM. (NB, figures for the male workers are weighted by a factor of three.)

BIBLIOGRAPHY

1. *Archives*

ACLNAI. Archivio del Comitato di liberazione nazionale alta Italia. Kept in Istituto nazionale per la storia della Resistenza e del movimento di liberazione in Italia. [Milan]

ACLNL. Archivio del Comitato di liberazione nazionale della Lombardia. Kept in Istituto nazionale per la storia della Resistenza e del movimento di liberazione in Italia. [Milan]

ACS. Archivio centrale di stato. Fondo Osservatori industriali (OI), Fondo Pubblica sicurezza (PS). [Rome]

AISRMO. Archivio dell'Istituto milanese per la storia della Resistenza e del movimento operaio. Kept in ISRMO. [Sesto San Giovanni]

APCI. Archivio del Partito comunista italiano. Kept in Istituto Gramsci. [Rome]

APMM. Archivio del personale Magneti Marelli. [Sesto San Giovanni]

ASM. Archivio di stato di Milano. Gabinetto del Prefettura (Gab. Pref.). [Milan]

2. *Newspapers*

L'assistenza sociale nell'industria (*ASNI*). Fascist publication.

Azione Sindacale. Publication of Milanese Fascist unions.

Il lavoro fascista. Fascist union publication.

Maternità ed infanzia. Bolletino ufficiale dell'Opera Nazionale per la protezione della Maternità e dell'Infanzia. Official publication of ONMI (Opera nazionale per la protezione della maternità e dell'infanzia).

L'organizzazione scientifica del lavoro. Official publication of ENIOS (Ente nazionale italiano per l'organizzazione scientifica del lavoro).

Il Popolo d'Italia. Fascist newspaper.

Il Popolo di Lombardia. Local newspaper.

Realtà. Publication of the Rotary Club of Milan.

Sprazzi e Bagliori (*SB*). Factory newspaper published by Magneti Marelli 1924–43.

Lo Stato Operaio. Communist antifascist publication. Published by political exiles in New York.

Unità. Newspaper of the Communist Party. Published clandestinely for most of the Fascist period.

3. *Published Books and Articles*

ACCORNERO, A., 'Dove cercare le origini del taylorismo e del fordismo', *Il Mulino*, 241 (1975).

AITKIN, H. G. J., *Taylorism at Watertown Arsenal: Scientific Management in Action* (Cambridge, Mass., 1960).

ALASIA, F., *Gaetano Invernizzi dirigente operaio* (Milan, 1976).

—— (ed.), *La vita di prima* (Milan, 1984).

ALLOISIO, M. and AJO, M., *La donna nel Socialismo Italiano tra cronaca e storia (1892–1978)* (Bari, 1978).

—— and BELTRAMI, G., *Volontarie della libertà* (Milan, 1981).

ANDREUCCI, F., 'Subversiveness and Anti-Fascism in Italy', in Samuel, R., *People's History and Socialist Theory* (London, 1981).

Anna Kuliscioff e l'età del riformismo (Atti del convegno di Milano, Dec. 1976) (Milan, 1977).

AQUARONE, A., 'La politica sindacale del fascismo', in A. Aquarone (ed.), *Il regime fascista* (Bologna, 1974).

ARIANO, A., CANAVESI, G. P., GAY, M., RIVA, N., SABATELLI, F., and TRABATTONI, L., 'I dormitori-convitti e i villaggi operai in una zona tessile del nord Milano', *Classe*, 14 (1977).

AUDIBERT, P., 'Nota sul gabinetto psico-tecnico della Miniera di Gennimari Ingurtosu', *Resoconti dell'Associazione Minieria Sarda*, 8 (1929).

BAGLIONI, G., 'La costruzione del paternalismo organico nel pensiero di un imprenditore d'eccezione: A. Rossi', *Studi di sociologia*, 3–4 (1971).

—— *L'ideologia della borghesia industriale nell'Italia liberale* (Turin, 1974).

BALLESTRERO, M. V., *Dalla tutela all parita: la legislazione italiana sul lavoro delle donne* (Bologna, 1979).

BARBAGLI, M., *Disoccupazione intellettuale e sistema scolastico in Italia, 1859–1973* (Bologna, 1974).

BEDAUX, C. E., *The Bedaux Efficiency Course for Industrial Application* (Grand Rapids, Mich., 1917).

BEECHEY, V., 'The Sexual Division of Labour and the Labour Process: A Critical Assessment of Braverman', in S. Wood (ed.), *The Degradation of Work? Skill, Deskilling and the Labour Process* (London, 1983).

BELL, D., *Sesto San Giovanni: Workers, Culture and Politics in an Italian Town, 1880–1922* (New Brunswick, NJ, 1986).

—— 'Worker Culture and Worker Politics: The Experience of an Italian Town, 1880–1915', *Social History*, 3 (1978).

—— 'Working-Class Culture and Fascism in an Italian Industrial Town, 1918–22', *Social History*, 1 (1984).

BELOTTI, V., CARBOGNIN, M., FELTRIN, P., MANTOVIN, P., PIVA, F.,

RAVANNE, F., and TATTARA, G., *I primi operai di Marghera: mercato, reclutamento, occupazione 1917–1940* (Venice, 1983).

BIANCHI, B., 'I tessili: lavoro, salute, conflitti', *Annali Feltrinelli* (1979–80).

BIANCHI, G. (ed.), *Dalla Resistenza: uomini, eventi, idee della lotta di Liberazione in provincia di Milano* (Milan, 1975).

BIFFOLI, C., and LUNGONELLI, M., 'Una classe operaia in formazione: i siderurgici di Portoferraio (1901–1905)', *Studi storici*, 1 (1985).

BIGAZZI, D., 'Management strategies in the Italian automobile industry', in J. Zeitlin and S. Tolliday (eds.), *Between Fordism and Flexibility: The International Automobile Industry and its Workers* (London, 1986).

—— 'Organizzazione del lavoro e razionalizzazione nella crisi del fascismo 1942–3', *Studi storici*, 2 (1978).

—— 'Gli operai della catena di montaggio: la Fiat 1922–1943', *Annali Feltrinelli* (1979–80).

—— ' "I più turbolenti della città: la composizione operaia all'Alfa Romeo', in G. Procacci (ed.), *Stato e classe operaia in Italia durante la prima guerra mondiale* (Milan, 1983).

—— *Il Portello: operai, tecnici e imprenditori all'Alfa Romeo 1906–1926* (Milan, 1988).

BINCHY, D. A., *Church and State in Fascist Italy* (Oxford, 1941).

BINEL, C., *Dall'Ansaldo alla Cogne: un'esempio di siderurgia integrale (1917–1945)* (Milan, 1985).

BOLDRINI, A., *Enciclopedia della Resistanza* (Bologna, 1980).

BONAZZI, G., 'Il taylorismo tra strumento del capitale e utopia tecnocratica (II)', *Economia del lavoro*, 2 (1972).

BORGOMANERI, L., *Due inverni, un'estate e la rossa primavera: le brigate Garibaldi a Milano e provincia 1943–1945* (Milan, 1985).

BOSISIO, L., 'Gli insediamenti industriali e l'urbanistica della transizione, 1880–1914', in G. Petrillo (ed.), *La città delle fabbriche: Sesto San Giovanni 1880–1945* (Milan, 1981).

—— 'Nascita e sviluppo del settore meccanico e siderurgico', in G. Petrillo (ed.), *La città delle fabbriche: Sesto San Giovanni 1880–1945* (Milan, 1981).

BOTTIGLIERI, B., 'Razionalizzazione del lavoro e salute operaia tra le due guerre: l'atteggiamento del sindacato e del governo', in M. L. Betri and A. Gigli Marchetti (eds.), *Salute e classi lavoratrici in Italia dall'Unità al fascismo* (Milan, 1982).

BRAMBILLA, G., 'Il movimento operaio nel 1943 a Milano', *Critica marxista* (1974).

BRAVERMAN, H., *Labor and Monopoly Capital: The Degradation of Work in the Twentieth Century* (New York, 1974).

BRAYBON, G., and SUMMERFIELD, P., *Out of the Cage: Women's Experience in Two World Wars* (London, 1987).

BRIDENTHAL, R., GROSSMAN, A., and KAPLAN, M. (eds.), *When Biology Became Destiny: Women in Weimar and Nazi Germany* (New York, 1984).

BRONZINI, G., 'Legislazione sociale e istituzioni corporativi', *Annali Feltrinelli* (1979–80).

BRUZZONE, A. M., and FARINA, R., *La Resistenza taciuta: dodici vite di partigiane piemontesi* (Milan, 1976).

BUJRA, J. M., 'Female Solidarity and the Sexual Division of Labour', in P. Caplan and J. M. Bujra (eds.), *Women United, Women Divided, Crosscultural Perspectives on Female Solidarity* (London, 1978).

BURCHARDT, L., 'Technischer Fortschritt und sozialer Wandel: Das Beispiel der Taylorismus in Deutschland', in W. Treue (ed.), *Deutsche Technikgeschichte* (Göttingen, 1977).

BUTTAFUOCO, A. R., *Le Mariuccine* (Milan, 1985).

CALDWELL, L., 'Reproducers of the Nation: Women and the Family in Fascist Policy', in D. Forgacs (ed.), *Rethinking Italian Fascism* (London, 1986).

CANDELORO, G., *Storia dell'Italia moderna*, ix: *Il fascismo e le sue guerre* (Milan, 1981).

CANNISTRARO, P. V., *Historical Dictionary of Fascist Italy* (Westport, Conn., 1982).

CAPLAN, J., 'Introduction to Female Sexuality in Fascist Ideology', *Feminist Review*, 1 (1979).

CASADIO, G., and FENATI, J. (eds.), *Le donne ravennati nell'antifascismo e nella Resistenza* (Ravenna, 1977).

CASALINI, Maria, 'Femminismo e socialismo in Anna Kuliscioff 1890–1907', *Italia contemporanea*, 143 (1981).

CASALINI, Mario, *La protezione e l'assicurazione della maternità e la difesa della prima infanzia* (Rome, 1934).

CASTLE, J., 'Factory Work for Women', in B. Lancaster and T. Mason (eds.), *Life and Work in a Twentieth Century City: The Experience of Coventry* (Coventry, 1986).

CASTRONOVO, V., *Giovanni Agnelli: la Fiat dal 1899 al 1945* (Turin, 1977).

CEREJA, F., 'La scuola e il mondo del lavoro: problemi dell'istruzione tecnica e professionale', *Annali Feltrinelli* (1979–80).

CHERUBINI, A., *Storia della previdenza sociale in Italia 1860–1890* (Rome, 1977).

CHIANESE, G., *Storia sociale della donna in Italia (1800–1980)* (Naples, 1980).

CLARK, M. N., *Antonio Gramsci and the Revolution that Failed* (New Haven, Conn., 1977).

COHEN, J., 'Was Italian Fascism a Developmental Dictatorship?', *Economic History Review* (1988).

CONSONNI, G., and TONON, G., 'Aspetti della questione urbana a Milano dal fascismo alla ricostruzione', *Classe*, 12 (1976).

——— 'Milano, classe e metropoli tra due economie di guerra', *Annali Feltrinelli* (1979–80).

COPLEY, F. B., *Frederick W. Taylor: Father of Scientific Management* (2 vols., New York, 1923).

CORNER, P., 'Fascist Agrarian Policy and the Italian Economy in the Interwar Years', in J. A. Davis (ed.), *Gramsci and Italy's Passive Revolution* (London, 1979).

COVINO, R., and GALLO, G., 'La forza lavoro della fabbrica d'armi di Terni durante la prima guerra mondiale', in G. Procacci (ed.), *Stato e classe operaia in Italia durante la prima guerra mondiale* (Milan, 1983).

CRESPI, P., *Capitale operaia: storie di vita raccolte tra le fabbriche di Sesto San Giovanni* (Milan, 1979).

DE BERNARDI, A., 'Operai, sindacati e regime negli anni venti: il caso di Milano', *Società e storia*, 40 (1988).

DE FELICE, R., *Mussolini il duce*, i: *Gli anni del consenso* (Turin, 1974).

DE GRAND, A., 'Women under Italian Fascism', *Historical Journal*, 4 (1976).

DE GRAZIA, V., *The Culture of Consent: Mass Organization of Leisure in Fascist Italy* (Cambridge, 1981).

—— 'Disciplina del lavoro e mediazione sociale sotto il regime fascista: le funzioni del dopolavoro nell'organizazzione del lavoro', *Annali Feltrinelli* (1979–80).

DELLA VALENTINA, G., 'Agricoltura e aspetti del rapporto città campagna in Lombardia tra crisi e seconda guerra mondiale', *Società e storia*, 20 (1983).

DETRIAGACHE, D., 'L'Italie fasciste e la répression de l'avortement', *Mélanges de l'École Française de Rome* (1980).

DE VECCHI, G., 'L'industrializzazione della fascia nord Milano e la localizzazione delle industrie a Sesto San Giovanni', and 'Origini e prima organizzazione del movimento operaio e democratico', both in G. Petrillo (ed.), *La città delle fabbriche: Sesto San Giovanni 1880–1945* (Milan, 1981).

DEVINAT, P., *Scientific Management in Europe*, International Labour Office, Studies and Reports, Series B, 17 (Geneva, 1927).

DOBB, M., *Wages* (Cambridge, 1960; first edition, 1928).

DODI OSNAGHI, L., 'Aspetti della condizione operaia e della nocività attraverso le riviste di medicina del lavoro', *Annali Feltrinelli* (1979–80).

DONVITO, A., and GARBARINI, G., 'Senz'altra formalità che il reciproco preavviso: le officine di Savigliano, 1904–1914', *Italia contemporanea*, 157 (1984).

ELLWOOD, D., 'From Re-education to the Selling of the Marshall Plan in Italy', in N. Pronay and K. Wilson (eds.), *The Political Re-education of Germany and her Allies after World War Two* (London, 1985).

ELLWOOD, D., *Italy 1943–45* (Leicester, 1985).

Fascio di combattimento di Sesto San Giovanni, *Sesto San Giovanni fascista industriale* (Milan, 1934).

FESTA, G., *L'Opera Nazionale per la protezione della Maternità e dell'Infanzia dalla sua fondazione* (Rome, 1962).

FIORENTINI, P., 'Ristrutturazione capitalistica e sfruttamento operaio in Italia negli anni '20', *Rivista storica del socialismo*, 30 (1967).

FORD, H. (in collaboration with S. Crowther), *My Life and My Work* (London, 1924; first published 1922).

GANAPINI, L. (ed.), '. . . *che tempi, però erano bei tempi* . . .': *la commissione interna della Magneti Marelli nella memoria dei suoi protagonisti* (Milan, 1986).

—— *Una città, la guerra: Milano 1939–1951* (Milan, 1988).

—— 'Lotte operaie Milano', in G. Quazza (ed.), *Operai e contadini nella crisi italiana del 1943–44* (Milan, 1974).

—— 'Milano, autunno 1944: la svolta delle lotte operaie', *Italia contemporanea* (1975).

GENCARELLI, E., 'Partito e classe a Milano negli scioperi del 1942–44', *Italia contemporanea*, 114 (1974).

GRAMSCI, A., *Americanismo e fordismo* (Turin, 1975; but written in the 1930s).

GRIFFEN, C., HOBSON, D., MACKINTOSH, S., and MCCABE, T., 'Women and Leisure', in J. Hargreaves (ed.), *Sport, Culture and Ideology* (London, 1982).

GROSSI, G., *La protezione e la difesa sociale della maternità in regime fascista* (Milan, 1930).

GUIDETTI SERRA, B., *Compagne* (2 vols., Turin, 1977).

GUIOTTO, L., *La fabbrica totale: paternalismo industriale e città sociale in Italia* (Milan, 1979).

GUIZZI, F., 'La Fabbrica Italiana Magneti Marelli', in G. Petrillo and A. Scalpelli (eds.), *La ricostruzione nella grande industria: strategia padronale e organismi di fabbrica nel Triangolo 1945–1948* (Bari, 1978).

HABER, S., *Efficiency and Uplift: Scientific Management in the Progressive Era* (Chicago, 1964).

HAREVEN, T. K., *Family Time and Industrial Time: The Relationship between the Family and Work in a New England Community* (Cambridge, 1982).

HIGGS, E., 'Women, Occupations and Work in the 19th Century Censuses', *History Workshop*, 23 (1987).

HOMBERG, H., 'Anfänge des Taylorsystems in Deutschland vor dem Ersten Weltkrieg: Eine Problemskizze unter besonderer Berücksichtigung der Arbeitskämpfe bei Bosch 1913', *Geschichter und Gesellschaft*, 2 (1978).

Il chi e? nella vita economica italiana (Milan, 1972).

JEMOLO, C. A. (trans. D. Moore), *Church and State in Italy, 1850–1950* (Oxford, 1960).

JOCTEAU, G. C., 'La magistratura del lavoro nello stato fascista: le contraversie collettive (1926–35)', in A. Aquarone (ed.), *Il regime fascista* (Bologna, 1974).

JONES, B., 'Destruction or Redistribution of Engineering Skills? The Case of Numerical Control', in S. Wood (ed.), *The Degradation of Work? Skill, Deskilling and the Labour Process* (London, 1983).

LAMBERTZ, J., 'Sexual Harassment in the Nineteenth Century English Cotton Industry', *History Workshop*, 19 (1985).

LANDES, D., *The Unbound Prometheus* (Cambridge, 1969).

LAYTON, E., 'The Diffusion of Scientific Management and Mass Production from the U.S. in the Twentieth Century', *Proceedings of the 16th International Congress in the History of Science* (Tokyo, 1974).

LEVI, F., 'Per una storia della fabbrica in Italia: una proposta di lavoro', *Rivista di storia contemporanea*, 4 (1977).

LEWIS, J., *Women in England, 1870–1950* (Brighton, 1984).

LICINI, S., 'Operai a Sesto San Giovanni: composizione e mobilità della forza lavoro alla Ercole Marelli (1929–1945)', *Storia in Lombardia*, 3 (1987).

LINCOLN CADIOLO, P., *Sesto San Giovanni dalle origini ad oggi* (Milan, 1964).

LITTERER, J. A., 'Systematic Management: The Search for Order and Integration', *Business History Review*, 35 (1981).

LIVI BACCI, M., *A History of Italian Fertility During the Last Two Centuries* (Princeton, NJ, 1977).

LYTTLETON, A., *The Seizure of Power: Fascism in Italy, 1919–29* (London, 1973).

MACCIOCCHI, M. A., *La donna nera: consenso femminile e fascismo* (Milan, 1976).

—— (ed.), *Les femmes e leurs maîtres* (Paris, 1978).

—— 'Female Sexuality in Fascist Ideology', *Feminist Review*, 1 (1979).

MAHER, V., 'Un mestiere da raccontare: sarte e sartine fra le due guerre', *Memoria*, 8 (1983).

MAIER, C. S., 'Between Taylorism and Technocracy: European Ideologies and the Vision of Industrial Productivity in the 1920s', *Journal of Contemporary History*, 2 (1970).

—— *Recasting Bourgeois Europe: Stabilization in France, Germany and Italy in the Decade after World War I* (Princeton, NJ, 1975).

MAIOCCHI, R., 'Scienza, industria e fascismo (1923–1939)', *Società e storia*, 2 (1978).

MEAKIN, W., *The New Industrial Revolution* (London, 1928).

MELDINI, P., *Spose e madri esemplari* (Rimini–Florence, 1975).

MELOGRANI, P., 'Benni Antonio Stefano', *Dizionario Biografico degli Italiani*, 8 (Rome, 1966).

—— *Gli industriali e Mussolini: rapporti tra Confindustria e Fascismo dal 1919 al 1929* (Milan, 1980).

MERLI, S., *Proletariato di fabbrica e capitalismo industriale* (Florence, 1972).

Milano tra guerra e dopoguerra (Milan, 1979).

MORRA, A., and SALA, P., 'In-migration, Population Turnover, Industrialization: Sesto San Giovanni 1900–1920', *Contribution of Italian Scholars. International Union for the Scientific Study of Population. XX General Conference* (Rome, 1985).

MUSSOLINI, B., *Discorso dell'ascensione: il regime fascista per la grandezza dell'Italia* (Rome–Milan, 1927).

Nei campi della gloria (Milan, 1924).

NOBLE, D., *America by Design: Science, Technology, and the Rise of Corporate Capitalism* (Oxford, 1977).

NOETHER, E. P., 'Italian Women and Fascism: A Re-evaluation', *Italian Quarterly*, 90 (1982).

OAKLEY, A., 'Interviewing women: A contradiction in terms', in H. Roberts, *Doing Feminist Research* (London, 1981).

ORTOLEVA, P., 'Henry, Heinrich, Enrico Ford: il fordismo come luogo comune nella cultura europea degli anni venti', introduction to H. Ford, *La mia vita e la mia opera*, trans. S. Benco (Milan, 1980).

PAPA, A., *La storia politica della radio in Italia* (2 vols., Naples, 1978).

PARKER, R., *The Subversive Stitch: Embroidery and the Making of the Feminine* (London, 1984).

PASSERINI, L., *Fascism and Popular Memory* (London, 1987).

—— 'Conoscenza storica e storia orale: sull'utilità e il danno delle fonti orali per la storia', in Passerini (ed.), *Storia orale: vita quotidiana e cultura materiale delle classi subalterne* (Turin, 1978).

—— 'Work Ideology and Consensus under Italian Fascism', *History Workshop*, 8 (1979).

—— 'Soggettività operaia e fascismo, indicazioni di ricerca dalle fonti orali', *Annali Feltrinelli* (1979–80).

—— 'Fonti orali e storia della classe operaia in regime fascista', in A. Agosti and G. Bravo (eds.), *Storia del movimento operaio, del socialismo e delle lotte sociali in Piemonte* (Bari, 1980).

—— 'Donne operaie e aborto nel Torino fascista', *Italia contemporanea*, 151–2 (1983).

PEDROCCO, G., *Fascismo e nuove technologie: l'organizzazione industriale da Giolitti a Mussolini* (Bologna, 1980).

PELAIA, M., 'Mestieri femminili e luoghi comuni: le domestiche a Roma a metà Ottocento', *Quaderni storici*, 68 (1988).

PERSON, H. S., 'The Bedaux System', *New Republic*, 24 Nov. 1937.

PESCE, G., *Senza tregua: la guerra dei GAP* (Milan, 1967).

PETRILLO, G., and VIGNATI, G. (eds.), *Città e fabbrica nella Resistenza, Sesto San Giovanni 1943–1945/documenti* (Sesto San Giovanni, 1975).

PHILIPS, A., and TAYLOR, B., 'Sex and Skill: Notes towards a Feminist Economics', *Feminist Review*, 6 (1980).

PICHIERRI, A., 'Diffusion and Crisis of Scientific Management in European Industry', in S. Giner and M. Scotford Archer (eds.), *Contemporary Europe: Social Structures and Cultural Patterns* (London, 1978).

PIERONI BORTOLOTTI, F., *Alle origini del movimento femminile, 1848–1892* (Turin, 1963).

—— *Socialismo e questione femminile in Italia' 1892–1922* (Milan, 1974).

POLLARD, J., *The Vatican and Italian Fascism 1929–1932: A Study in Conflict* (Cambridge, 1985).

PORTELLI, A., 'Peculiarities of Oral History', *History Workshop*, 12 (1981).

POZZONI, E., *Commemorazione di Antonio Stefano Benni* (Milan, 1959).

PROCACCI, G., *La lotta di classe in Italia agli inizi del secolo XX* (Rome, 1972).

PUCCINI, S., 'Condizione della donna e questione femminile (1892–1922)', in G. Ascoli (ed.), *La questione femminile in Italia dal '900 ad oggi* (Milan, 1979).

RAVERA, C., *La donna italiana dal primo al secondo Risorgimento* (Rome, 1951).

REY, G. M., 'Una sintesi dell'economia italiana durante il fascismo', in G. Toniolo (ed.), *L'economia italiana 1861–1940* (Rome–Bari, 1978).

ROMANO, R., 'Gli industriali e la prevenzione degli infortuni sul lavoro (1894–1914)', in M. L. Betri and A. Gigli Marchetti (eds.), *Salute e classi lavoratrici in Italia dall'Unità al fascismo* (Milan, 1982).

ROSSI, C., 'Antonio Stefano Benni: un capo venuto dalla gavetta', in Rossi, *Personaggi di ieri e di oggi* (Milan, 1960).

RUGAFIORI, P., 'Occupazione e composizione operaia all'Ansaldo', in G. Procacci (ed.), *Stato e classe operaia in Italia durante la prima guerra mondiale* (Milan, 1983).

—— *Uomini, macchine, capitali: l'Ansaldo durante il fascismo 1922–1945* (Milan, 1981).

RULFI, G., 'Un centro industriale lombardo: Sesto San Giovanni', *Rivista geografica italiana* (1955).

RUMI, G., 'Padre Gemelli e l'Università Cattolica', in *Modernismo, fascismo, comunismo* (Bologna, 1972).

SABBATUCCI SEVERINI, P., and TRENTO, A., 'Alcuni cenni sul mercato di lavoro durante il fascismo', *Quaderni storici*, 29–30 (1975).

SALVEMINI, G., *Under the Axe of Fascism* (London, 1936).

SANTARELLI, E., 'Il fascismo e le ideologie antifemministe', *Problemi del socialismo* (1976).

SAPELLI, G., 'Formazione della forza lavoro e psicotecnica nell'Italia tra le due guerre mondiali', *Quaderni di sociologia*, 1 (1977).

—— *Organizzazione, lavoro e innovazione industriale nell'Italia tra le due guerre* (Turin, 1978).

—— 'Per la storia del sindacalismo fascista: tra controllo sociale e conflitto di classe', *Studi storici* (1978).

SARACENO, C., 'La famiglia operaia sotto il fascismo', *Annali Feltrinelli* (1979–80).

—— 'Percorsi di vita femminile nella classe operaia: tra famiglia e lavoro durante il fascismo', *Memoria*, 2 (1981).

SARTI, R., *Fascism and the Industrial Leadership, 1919–40* (Berkeley, 1971).

SCALPELLI, A., *Scioperi e guerriglia in Val Padano* (Urbino, 1972).

SCARAMUZZA, E., 'Professioni intellettuali e fascismo: l'ambivalenza dell'Alleanza muliebre culturale italiana', *Italia contemporanea*, 151–2 (1983).

SCHWARZENBERG, C., *Il sindacalismo fascista* (Milan, 1973).

SCOPPOLA, P., *La chiesa e il fascismo: documenti e interpretazioni* (Bari, 1971).

SERENI, E., *Cln, il comitato di liberazione nazionale della Lombardia al lavoro nella cospirazione, nell'insurrezione, nella ricostruzione* (Milan, 1945).

Sesto San Giovanni nella Resistenza (Sesto San Giovanni, 1973).

SPRIANO, P., *L'occupazione della fabbriche: settembre 1920* (Turin, 1964).

—— *Storia del Partito Comunista Italiano*, v: *La Resistenza, Togliatti e il partito nuovo* (Turin, 1976).

TADDEI, B., *Donne processate dal Tribunale speciale 1927–43* (Verona, 1969).

TAGLIACARNE, G., 'L'occupazione delle donne nelle fabbriche e negli uffici durante gli ultimi cinquant'anni e la diminuzione delle nascite', *Giornale degli economisti e rivista di statistica* (Dec. 1934).

TANNENBAUM, E., *The Fascist Experience: Italian Society and Culture 1922–1945* (New York, 1972).

TAYLOR, F. W., *Scientific Management: Shop Management, the Principles of Scientific Management, Testimony before the Special House Committee* (Westport, Conn., 1972; first edition published 1911).

THOMPSON, C. B., 'Scientific Management in Practice', *Quarterly Journal of Economics* (1915).

THOMPSON, P., *The Voice of the Past: Oral History* (Oxford, 1978).

TOGLIATTI, P., *Lezioni sul fascismo* (Rome, 1970).

TONIOLO, G., *L'economia italiana 1861–1940* (Rome–Bari, 1978).

TRAUB, R., 'Lenin and Taylor: The Fate of "Scientific Management" in the (Early) Soviet Union', *Telos*, 37 (1978).

TREMPÉ, R., *Les Mineurs de Carmaux, 1848–1914*, 2 vols. (Paris, 1971).

—— 'Pour une meilleure connaissance de la classe ouvrière: l'utilisation des archives d'entreprises; le fichier du personnel', *Mélanges d'histoire sociale offerts à Jean Maitron* (Paris, 1978).

TREVES, A., *Le migrazioni interne nell'Italia Fascista* (Turin, 1976).

VACCARI, I., *La donna nel ventennio fascista 1919–43* (Milan, 1978).
VAUDAGNA, M., 'L'américanisme e le management scientifique dans l'Italie des années 1920', *Recherches*, 32–3 (1978).
VIGNATI, G., 'Un'opposizione dura a morire: l'antifascismo nelle fabbriche e nella città', in G. Petrillo (ed.), *La città delle fabbriche: Sesto San Giovanni 1880–1945* (Milan, 1981).
—— (ed.), *I ribelli al governo della città: Sesto San Giovanni 1944–1946* (Milan, 1988).
WILLIAMS, G. A., *Proletarian Order: Antonio Gramsci and the Origins of Communism in Italy 1911–1921* (London, 1975).
WILLSON, P., ' "The Golden Factory": Industrial Health and Scientific Management in an Italian Light Engineering Firm; the Magneti Marelli in the Fascist Period', in P. Weindling (ed.), *The Social History of Occupational Health* (London, 1985).
—— 'Lavoro femminile, ideologia del lavoro e relazioni in fabbrica', in L. Ganapini (ed.), '*. . . che tempi, però erano bei tempi . . .*': *la commissione interna della Magneti Marelli nella memoria dei suoi protagonisti* (Milan, 1986).
WOOD, S. (ed.), *The Degradation of Work? Skill, Deskilling and the Labour Process* (London, 1982).
ZAVALLONI, R., *Padre Gemelli educatore sociale* (Milan, 1960).
ZIMMICK, M., 'Clerical Work for Women, 1850–1914', in A. John (ed.), *Unequal Opportunities* (London, 1986).

4. *Unpublished Secondary Sources*

ABENDSTERN, M., 'Expression and Control: A Study of Working Class Leisure and Gender 1918–1939. A Case Study of Rochdale Using Oral History Methods', Ph.D. thesis (Essex, 1986).
AMOROSO, A., 'I gruppi di difesa della donna a Milano', paper given at conference on 'Donna lombarda 1860–1945' (Milan, 1989).
DALMASSO, E., 'Milan, capitale économique de l'Italie', thèse de doctorat (Aix–Marseille, 1970).
GIUNTINI, S., 'La donna e lo sport in Lombardia durante il fascismo', paper given at conference on 'Donna lombarda 1860–1945' (Milan, 1989).
LONGARETTI, L., 'Il fascio di Sesto San Giovanni: storia delle camicie nere in un centro industriale 1919–29', tesi di laurea (Milan, 1983–4).
MONGIO, S., and RIMOLA, M., 'Per una definizione del rapporto donne-istituzione: il caso ONMI dalla fondazione al 1935', tesi di laurea (Trento, 1977–8).
TINTORRI, F., 'L'OND a Milano (1923–1939)', paper given at conference on 'Industria, fabbrica, territorio nella Lombardia degli anni '30' (Milan, 1984).

INDEX